PROGRESS IN FOOD SCIENCE AND TECHNOLOGY

# PHYTOSTEROLS

# FOOD SOURCES, FUNCTIONS AND HEALTH BENEFITS

# PROGRESS IN FOOD SCIENCE AND TECHNOLOGY

Additional books in this series can be found on Nova's website under the Series tab.

Additional e-books in this series can be found on Nova's website under the e-book tab.

PROGRESS IN FOOD SCIENCE AND TECHNOLOGY

# PHYTOSTEROLS

# FOOD SOURCES, FUNCTIONS AND HEALTH BENEFITS

DEANNA GARNER
EDITOR

nova publishers
New York

**Library of Congress Cataloging-in-Publication Data**

Library of Congress Control Number: 2015946403

ISBN: 978-1-63483-477-3

*Published by Nova Science Publishers, Inc. † New York*

# CONTENTS

# PREFACE

The consumption of phytosterols (plant sterols and/or stanols) can reduce cardiovascular risk by reducing hypercholesterolemia. In addition to reducing intestinal cholesterol absorption, phytoserols (PS), which are structurally similar to cholesterol, have other potential beneficial effects in humans. There are a wide range of plants that have phytosterols in their composition. In this book, the food sources of phytosterols are explored. The authors evaluate whether a decrease in carotenoids associated with the consumption of a therapeutic dose of PS alters oxidative stress parameters. Furthermore, the characterization of oils by phytosterols analysis was studied and is examined in this book. An overview of phytosterols in olive oil is given as well, particularly, the use of HPLC and GC coupled to various detection systems to isolate sterols from olive oil in order to focus on the occurrence of phytosterols and their role in human health.(

Chapter 1 – Cardiovascular diseases are the primary cause of mortality and morbidity in the world and are associated with clearly higher levels of total cholesterol and low density lipoprotein. Phytosterols are cholesterol-like molecules found in all vegetal-derived foods items, but consumption of these compounds has a cardioprotective effect by lowering blood cholesterol levels. The efficacy, effectiveness and safety of phytosterols as a cholesterol-lowering therapy has already been corroborated in numerous clinical trials and meta-analyses by a dose of 2 g/day. However, the estimated average intake of naturally occurring plant sterols in the diet constitutes only about 10-20% of the recommended daily intake of 2 g and it is therefore necessary to ingest phytosterols-enriched foods or phytosterols supplements in order to reach the optimal dose.

Currently, various products enriched with phytosterols, including margarine, yogurt, milk, juice, cereal bars or nutraceuticals (tablets or capsules) are available on the market. With the increasing growth of this market it is important to alert for the risks inherent to the indiscriminate consumption of this type of products, and also to control the quality and their possible adulteration, in order to ensure a high level of protection of public health.

Qualitative and quantitative determination of phytosterols is well-documented in various food matrices as in vegetables, oil, fruit, cereal, bread, margarine and most recently, in milk and yoghurt. However for nutraceutical tablet or capsule forms there are no published methods. For this reason, it is essential to develop methodologies for the analysis of phytosterols in different food matrices, since each matrix is unique.

It seemed important to address the quality and the safety of phytosterol nutraceutical forms, contributing to the development of cholesterol-lowering products safety.

Thus, in the present chapter, a GC-MS chromatographic methodology is developed, optimized and validated for the determination of phytosterols in nutraceuticals. The application of the referred methodology to a nutraceutical formulation, shows that five phytosterols were present, namely campesterol, campestanol, stigmasterol, β-sitostanol and β-sitosterol. The application of the developed methodology in quality control is also discussed.

Chapter 2 – A wide range of plants have phytosterols in their composition, both strerols and stanols, being however the stanols present in considerably smaller quantities when compared to sterols. Among the most abundant plant sterols are sitosterol, campesterol and stigmasterol. As to plant stanols, they are present in the diet in small amounts, being one of the most important the sitostanol.

Some food sources that are rich in phytosterols include for example: fruits, vegetables, nuts, seeds, cereals, legumes and vegetable oils. It is therefore important to know what the food sources of these bioactive compounds are and what amounts are present. This is one of the goals of the present chapter, which aims to revise the most recent scientific literature to look for updates regarding the food sources of phytosterols, as a complement to the information more generally available.

Both stanols and sterols are essential components of plant cell membranes and structurally resemble cholesterol. In fact, cholesterol is also a sterol, but, unlike plant sterols, it is predominantly of animal origin, being synthesized in the human liver. Because the bioactive effects of phytosterols are much associated to their chemical structure, also this aspect is addressed in the present chapter, by presenting the chemistry of the different molecules of the phytosterols.

Chapter 3 – Refining of oils allows the elimination of color, odor and flavor that make crude oils unacceptable to consumers. In addition, chemical compounds that might be toxic or those that decrease the oil stability are removed in the refining process. Unfortunately, other components including phytosterols, of well-known bioactivity, are removed in part along with the undesired compounds and are ultimately concentrated in the refining by-products. Therefore, re-use and exploitation of by-products as an attractive source of bioactive compounds is gaining much interest. In this chapter, by-products from both chemical and physical refining used as sources of phytosterols, namely soapstocks and deodorization distillates, will be discussed. Soapstocks consist mainly of sodium salts produced from free fatty acids and also contain neutral oil and gums. Deodorization distillates include a considerable number of bioactive compounds, among them phytosterols, that make them excellent starting materials for obtaining novel food products of high added value. The extraction, isolation and purification of phytosterols from these raw materials using diverse technologies, including molecular distillation, extraction with supercritical carbon dioxide, hydrolysis of esters and crystallization and re-crystallization of adducts, will be revised.

Chapter 4 – Recent interest in phytosterols stems from their potential protective role, through ingestion of plant foods and oils. This latter approach in turn, results in reducing of serum total cholesterol, and low density lipoprotein (LDL) which may delay or/and prevent occurrence of chronic diseases like CHD and stroke. Many plant foods and vegetable oils contain a variety of phytochemicals with several properties related to individual compounds such as phytosterols which are present in low quantities. For the isolation of phytosterols, high-performance liquid chromatography (HPLC) as well as gas chromatography (GC) coupled with either an ultraviolet/visible (UV/Vis), photodiode array (PDA) and a flame ionization detector (FID) in GC have been widely used. Moreover, mass spectrometry (MS) detector with different analyzers such as ion trap (IT), time-of-flight (Q) TOF, has been hyphenated for identification of phytosterols from diverse vegetable oils. In this sense, the use of HPLC and GC coupled to various detection systems to isolate sterols from olive oil has been reviewed in this chapter. This review is devoted to a short discussion of the occurrence of phytosterols, their role in human health, and focuses on a detailed presentation of the analytical methods, concluding with the advantages of analytical methods employed so far.

Chapter 5 – The consumption of phytosterols (plant sterols and/or stanols) can reduce cardiovascular risk, by reducing hypercholesterolemia. Serum cholesterol concentrations can be determined by the traditional lipid profile parameters (total cholesterol, LDL-C, HDL-C, triglyceride), and in addition, the cholesterol metabolism can also be studied by the quantification of the intestinal cholesterol absorption and endogenous cholesterol synthesis markers. The disruption of the cholesterol homeostasis between absorption and synthesis of cholesterol is related to a positive history of several cardiovascular diseases and will be focused in this chapter.

Sitosterol and campesterol are the major sources of phytosterols in diet, and are therefore the main biomarkers used to study the exogenous cholesterol absorption, while desmosterol and lathosterol, the late precursors of the cholesterol biosynthetic pathway, are the most common biomarkers used to quantify cholesterol synthesis, whether in diet/lifestyle interventions or drug therapy monitoring.

There is evidence that several food matrices enriched with phytosterols, or supplements, are effective in lowering serum LDL-C concentrations, when used alone as part of the diet or when used as an adjuvant to ongoing statin therapy. In addition, the market for functional foods and dietary supplements with a health claim (for lipid-lowering) is expanding rapidly worldwide, and consequently an increasing number of persons will use these products and combine them with their prescribed drugs. Therefore, several studies have investigated the effect on lipids and also on serum cholesterol metabolism markers of a variety of phytosterol-enriched food matrices in hypercholesterolemic patients and in statin-treated individuals.

In addition to the hypocholesterolemic effect of the phytosterol-enriched food products, it is considered that in the particular case of the fermented dairy (low-fat) products, there are other benefits, such as nutritional, physiological and immunological positive overall effects on health, and in particular, in intestinal health.

In sum, this chapter will include a description of phytosterol-enriched food matrices studied/used to lower cholesterol levels and developments and trends in the field will be discussed. Studies on the use of these functional foods in hypercholesterolemic patients, on and off drug therapy, will be referenced. The use of the cholesterol metabolism markers will be highlighted in regard to the study of cardiovascular diseases, and the potential role of phytosterols will be discussed in disease prevention and health promotion.

Chapter 6 – *Background:* The article presents selected plasmatic cholesterol precursors (Lathosterol – Lat, Desmosterol – Des) and plasmatic phytosterols (Campesterol—Cam, Sitosterol—Sit) in relation to the parameters of cholesterol metabolism in 300 children patients (8-18 yrs) with genetically proved heterozygous familial hypercholesterolemias (HFH) and other hypercholesterolemias in order to obtain practical experience with the possible differential and monitoring use of these parameters in dietary or drug treatment. The authors discuss their 20 years of experience with the use of these seleceted non-cholesterol sterols (NCS) in diagnosing, differentiation, and dietary & drug treatment.

*Methods:* The plasma NCS values were determined using GC/MS; plasma lipid profile was established by routine clinical biochemistry methods; and molecular genetic analyses were conducted by PCR-SSCP and PCR-DGGE methods.

*Results:* The article indicates reference values of NCS for genetically proved healthy children of different ages. Patients with the so-called 'alimentary hypercholesterolemia' possess normal values of NCS. Children patients with HFH detected by positive selective screening (5th and 13th year of life) and arriving at the lipid outpatient clinic had usually been put on a low cholesterol diet (LCD) by pediatricians and had elevated levels of Lat and Des and compensatorily elevated Cam and Sit. Administration of simvastatin in sufficient therapeutic doses reduced Lat or Des ($p > 0.001$). Compensatory elevation of Cam and Sit persists in a few patients. Ezetimibe application decreases efficiently Cam and Sit rather than Lat and Des. Combined application of simvastatin with ezetimibe is the most efficient not only in the decrease in the plasma LDL-cholesterol, but also in decrease in Lat, Des as well as Cam and Sit. The diagnostic and therapeutic monitoring in patients with different types of hypercholesterolemia is also given in the graphic form.

*Conclusions:* Extending the laboratory monitoring profile by precursors of cholesterol synthesis (Lat, Des) and cholesterol absorption (Cam, Sit) improves the differential diagnosis power for different hypercholesterolemias in children and makes monitoring and treatment of children and adolescents suffering from HFH more precise and effective.

Chapter 7 – Early development of cardiovascular diseases is clearly associated with high plasma or serum cholesterol concentrations. Cholesterol homeostasis is reflected by the balance between cholesterol absorption *vs.* cholesterol synthesis. The ratios to serum cholesterol of sterol precursors reflect synthesis rate of cholesterol, while those of phytosterols reflect cholesterol absorption. A study with 95 volunteer healthy students from Coimbra University was developed in order to determine the prevalence of surrogate markers for cholesterol synthesis and absorption. Blood was collected and centrifuged and the serum was submitted to alkaline saponification, followed by *n*-hexane extraction and subsequently evaporation to dryness under a nitrogen stream. The dry residue was derivatized and the cholesterol and non-cholesterol sterols were determined by Gas Chromatography-Mass Spectrometry. Obtained results showed two relevant facts. Firstly, total cholesterol concentrations in the study population were quite high (21% of analyzed students with > 200 mg/dL), given the age group (19 to 24 years). Secondly, the cholesterol absorption markers (phytosterols) were found in concentrations higher than the markers of synthesis.

Obtained results can become an important contribution for developing better treatments to maintain healthy cholesterol concentrations, either for the elaboration of dietetic plans, namely for Mediterranean young people.

Chapter 8 – *Background:* Phytosterols (PS), which are structurally similar to cholesterol, reduce intestinal cholesterol absorption and have other potentially beneficial effects in humans. Based on recent meta-analyses, a PS intake of 2 g/day lowers LDLc by an average of 8-10%, though this effect varies depending on intrinsic and extrinsic factors particular to the individual. We have assessed whether baseline cholesterol concentration, amount of cholesterol in the diet, Apolipoprotein (Apo) E genotype, and presence of metabolic syndrome (MetS) influence the effectiveness of PS. Furthermore, the authors have evaluated whether a decrease in carotenoids associated with consumption of a therapeutic dose of PS alters oxidative stress parameters.

*Material and methods:* This dietary intervention study was performed in moderate hypercholesterolemic and normocholesterolemic subjects. The study was of a parallel, randomized design with three arms, and was conducted over a period of 6 months. The first arm was performed with 52 volunteers from the general population who continued their usual diet plus 2 g PS/day administered in 500 ml of low fat milk for 3 months. In the other two arms, 75 hypercholesterolemic individuals were administered a healthy diet for 3 months in order to standardize dietary habits. After this adjustment, subjects were divided into two intervention groups: one group followed a healthy diet with 2 g of PS/day for 3 months, and the other group followed the healthy diet without PS enrichment. Anthropometric parameters, lipid profile, Apo E genotype, oxidative stress parameters (susceptibility to oxidation in LDL particles, total antioxidant capacity and lipid peroxidation) and concentrations of serum carotenoids and PS (β-sitosterol and campesterol) were measured at the beginning and end of the study.

*Results:* Three-month dietary supplementation with PS produced a significant decrease in blood cholesterol (5.3% in total cholesterol and 6.5% in LDLc) in the general population without dietary restrictions. The reduction of LDLc was more pronounced in subjects with higher baseline concentrations of cholesterol. The beneficial effect of PS on the lipid profile of mildly hypercholesterolemic patients was not affected by the amount of saturated fat or cholesterol in their diet, nor was it mediated by the genotype of the Apo E. PS therapy appeared to be of little value to MetS patients, probably because of the reduced intestinal absorption of cholesterol. On the other hand, consumption of PS did not reduce serum total antioxidant capacity or increase oxidative stress.

*Conclusions:* Our results demonstrate that the efficacy of PS does not depend on the composition of macronutrients in the diet or the apolipoprotein E genotype; on the contrary, it is subject to baseline levels of LDL cholesterol. Moreover, PS therapy appears to be of little value to MetS patients.

Chapter 9 – Phytosterols are lipid compounds present in eukaryotic cells of plant species. The intracellular synthesis of phytosterols part of a complex of reactions starting from terpenoid alcohols, thus forming the structure of squalene. The group of tetracyclic fused rings has a flexible side chain whose structural changes between sterols involve differences in this chain. The phytosterols include several different combinations of atoms resulting in different possible structures, which they were identified in several plant oils. The β-sitosterol is the main phytosterols, and we can not underestimating occurrence of stigmasterol and campesterol. Phytosterols may appear to genuinely and mainly as free and conjugated steroids (FS, ES, SG and ASG). However, the profile of steroids, or the types of steroids and their relative proportions are not equal, thus making their fingerprint of identification. These compounds have been highlighted because of the chemical and pharmaceutical applications, especially with medical-therapeutic purposes. Another function of phytosterols, and not least important, is their composition used in the characterization of oils and the fingerprint of the same. The characterization of oils by phytosterols analysis has been studied in order to detect the adulteration of noble oils. Various methods have been used in the isolation of steroids and subsequent identification and quantification of the same.

Chapter 10 – Plant sterols (PS), including phytosterols and phytostanols, are bioactive compounds which when consumed have several beneficial effects on the human body, including antiinflammatory and anticancer activities. In addition, a daily intake of 1.5-3 g PS results in an 8-10% reduction in LDL-cholesterol levels. Previous studies in northern

European countries have reported PS intakes from natural and PS-enriched sources, though no recent studies have been made in the Mediterranean area, where the diet is known to be naturally rich in PS-containing foodstuffs (e.g., olive oil, nuts, etc.). Therefore, the aim of the present study was to describe the common PS sources and daily intakes of PS in a resident population (n = 75, 50.7% men, age = 20-68 years) in the Valencian Community (a region located on the Spanish Mediterranean coast) between November 2013 and May 2014, and to compare it with data collected in the year 2005 from a similar type of survey conducted among residents of the same region (n = 70, 47.1% men, age = 20-77 years). Intake was estimated by using a food frequency questionnaire considering naturally PS-containing food products from the normal diet. The results showed total PS intake to be similar in both study periods (274 mg/day in 2013-2014 versus 282 mg/day in 2005). Although the predominant PS content in natural sources corresponded to oils followed by nuts and legumes, on considering their frequency of consumption, the main PS sources were seen to be legumes > vegetables > oils. In sum, the recorded PS intake falls within the range reported by other studies in European populations (211-325 mg/day).

In: Phytosterols
Editor: Deanna Garner

ISBN: 978-1-63483-477-3


*Chapter 1*

# PHYTOSTEROLS IN NUTRACEUTICALS

***Patrícia Vicente[1], Maria da Conceição Castilho[1] and Fernando Ramos[1,2,*]***

[1]CEF - Center for Pharmaceutical Studies
[2]CNC - Center for Neuroscience and Cell Biology
Health Sciences Campus, Faculty of Pharmacy, University of Coimbra, Azinhaga de Santa Comba, Coimbra, Portugal

## ABSTRACT

Cardiovascular diseases are the primary cause of mortality and morbidity in the world and are associated with clearly higher levels of total cholesterol and low density lipoprotein. Phytosterols are cholesterol-like molecules found in all vegetal-derived foods items, but consumption of these compounds has a cardioprotective effect by lowering blood cholesterol levels. The efficacy, effectiveness and safety of phytosterols as a cholesterol-lowering therapy has already been corroborated in numerous clinical trials and meta-analyses by a dose of 2 g/day. However, the estimated average intake of naturally occurring plant sterols in the diet constitutes only about 10-20% of the recommended daily intake of 2 g and it is therefore necessary to ingest phytosterols-enriched foods or phytosterols supplements in order to reach the optimal dose.

Currently, various products enriched with phytosterols, including margarine, yogurt, milk, juice, cereal bars or nutraceuticals (tablets or capsules) are available on the market. With the increasing growth of this market it is important to alert for the risks inherent to the indiscriminate consumption of this type of products, and also to control the quality and their possible adulteration, in order to ensure a high level of protection of public health.

Qualitative and quantitative determination of phytosterols is well-documented in various food matrices as in vegetables, oil, fruit, cereal, bread, margarine and most recently, in milk and yoghurt. However for nutraceutical tablet or capsule forms there are no published methods. For this reason, it is essential to develop methodologies for the analysis of phytosterols in different food matrices, since each matrix is unique.

---

* Corresponding author: framos@ff.uc.pt.

It seemed important to address the quality and the safety of phytosterol nutraceutical forms, contributing to the development of cholesterol-lowering products safety.

Thus, in the present chapter, a GC-MS chromatographic methodology is developed, optimized and validated for the determination of phytosterols in nutraceuticals. The application of the referred methodology to a nutraceutical formulation, shows that five phytosterols were present, namely campesterol, campestanol, stigmasterol, β-sitostanol and β-sitosterol. The application of the developed methodology in quality control is also discussed.

## 1. Introduction

Food habits are a key factor in the health of individuals and the general population. An unbalanced diet low in essential nutrients may favor the onset of health problems such as obesity, dyslipidemia, hypertension, diabetes and high cholesterol.

Cholesterol is the most abundant sterol in humans, assuming a fundamental role in the body because it is a constituent of all cell membranes and plasma lipoproteins, especially low density lipoprotein (LDL). However, high levels of cholesterol lead to the early development of cardiovascular disease, which currently represents a major cause of mortality and morbidity in developed countries (Saraiva et al., 2011a; Myrie et al., 2012; Mannarino et al., 2014; Chen et al., 2014).

All strategies aimed at reducing cholesterol levels are also relevant for improving quality of life. The identification of natural compounds that assist in the control and prevention of cardiovascular risk has been increasingly studied, resulting in the discovery of sterols and plant stanols, commonly known as phytosterols.

Phytosterols are naturally occurring components in plants that carry out different biological functions similar to animal cholesterol because they are responsible for the fluidity and permeability of cell membranes (Trautwein and Demonty, 2007). These molecules have a great impact on human health by reducing cholesterol. Several clinical studies have demonstrated that a daily intake of 1.5 - 3 g of phytosterols reduced LDL-cholesterol by 10 to 15% (Katan et al., 2003), thus decreasing the formation of atherosclerosis plaques (Trautwein and Demonty, 2007). Phytosterols are also known to have anti-inflammatory (Li et al., 2007; He et al., 2013.), antiatherogenic (Trautwein and Demonty, 2007; He et al., 2013), antioxidant (Berger et al., 2004; Li et al., 2007) and anticancer (Awad and Fink, 2000; Lagarda et al., 2006; Li et al., 2007) properties.

Although phytosterols are consumed daily through foods naturally containing these substances, the values are not significant for cholesterol reduction (Tasan et al., 2006). The production of phytosterols therefore represents a business opportunity in both the food and pharmaceutical industries for the development of functional foods and nutraceuticals, respectively.

Advances in knowledge of both the disease process and healthy food components present new ways for pharmaceutical development and/or dietary strategies to stop the progression of vascular disease (Badimon et al., 2010). The growing awareness of the relationship between diet and health has stimulated growth in functional foods and nutraceuticals with the promise of health benefits for consumers (Hobbs et al., 2014).

During the last decades, numerous hypolipidemic nutraceuticals and functional foods have been discovered from epidemiological and intervention studies (Chen et al., 2014). It is therefore important to address the issue of the quality and safety of nutraceuticals with phytosterols to contribute to the development of safe cholesterol-lowering products.

In this book chapter, a description of phytosterols in food is presented (nomenclature and chemical structure, occurrence, properties, metabolism and mechanism), followed by a more detailed analysis of phytosterol enriched nutraceuticals (safety use, legislation, methods of analysis). The chapter concludes with some future prospects for improvement in the sector.

# 2. PHYTOSTEROLS IN FOOD

## 2.1. Nomenclature and Chemical Structure

Phytosterols are natural plant constituents essential for performing several cell functions, such as regulation of the fluidity and permeability of cell membranes. Phytosterols also serve as precursors of biogenic compounds involved in plant growth (Weingartner et al., 2009). They are structurally very similar to animal cholesterol, presenting a steroid nucleus, a 3β-hydroxyl group and a double bond between carbons 5 and 6. The main differences are in the alkyl side chain, through the presence or absence of a methyl or ethyl group, and the position of the double bond (Abidi, 2001; Lutjohann, 2004; Lagarda et al., 2006; Trautwein and Demonty, 2007). Phytosterols can be divided into three groups: 4-desmethyl, 4α-monomethyl and 4,4-dimethyl, according to their structure and biosynthesis. The 4-desmethylsterols are the group that assumes greater importance and may exist as free sterols (as alcohol) or may establish ester bonds with fatty acids through the 3-carbon hydroxyl or, more rarely, with hydroxycinnamic acids or ether connections with sugars (Clifton, 2004; Lagarda et al., 2006).

The most representative plant sterols, which are β-sitosterol (24α-ethylcholest-5en-3β-ol), campesterol (24α-methylcholest-5en-3β-ol) and stigmasterol (524α-ethylcholest-5,22-en-3β-ol), belong to the 4-desmethyl group. These plant sterols represent approximately 45-95%, 30% and 25%, respectively, of the phytosterols present in plants (Lutjohann, 2004; Woodgate et al., 2006; FAO, 2008). Stanols, which can be produced by hydrogenation of sterols (FAO, 2008) are present in saturated form and with no double bonds. The compounds of this subgroup are in lower abundance in nature, the most common being β-sitostanol and campestanol (Katan et al., 2003; Woodgate et al., 2006).

Phytosterols are insoluble in water but soluble in acetone and ethyl acetate (FAO, 2008). They have very high melting points (Srinivasan et al., 2008) and present as lipidic crystals at temperatures common to most foods. To minimize crystallization, they are esterified with unsaturated fatty acids to increase solubility in lipids (Srinivasan et al., 2008). The esterified phytosterols are liquid or semi-liquid compounds, which have physical and chemical characteristics comparable to edible fats and oils, allowing their addition to various foods (FAO, 2008). Phytosterols are stable compounds at room temperature but can be oxidized at elevated temperatures (above 100°C) (FAO, 2008). Stanols are more resistant to oxidation, but their application has been limited by presenting poor oil solubility and insolubility in water (He et al., 2013).

## 2.2. Occurrence in Foods

The most important natural sources of phytosterols in the human diet are margarines and oils, although they are also present in a variety of seeds, vegetables and unrefined vegetable oils (Piironen et al., 2000; Lagarda et al., 2006; Trautwein and Demonty, 2007) (Table 1).

Cereals, despite having a smaller amount of sterol compared to oils, contribute most significantly to the human diet because they are more regularly consumed (Lagarda et al., 2006). Analysis of the phytosterol levels in different grains found that rye is a better source than oats (Piironen et al., 2000). Fruits and vegetables are not the best natural sources of phytosterols compared to oils and cereals, but their contribution to the daily dietary intake is relevant (Piironen et al., 2000).

**Table 1. Phytosterol content in selected foods[a]**

| **Food** | | **Phytosterols (mg/100 g edible portion)** |
|---|---|---|
| **Vegetable Oils** | Rice Bran Oil | 1190 |
| | Corn oil | 968 |
| | Sesame oil | 865 |
| | Sunflower oil | 444 |
| | Soy oil | 250 |
| | Olive oil | 221 |
| | Peanut oil | 207 |
| **Seeds and Nuts** | Sesame seeds | 404 |
| | Sunflower seeds | 322 |
| | Linseeds | 213 |
| | Almonds | 143 – 208 |
| | Peanuts | 116 – 220 |
| | Hazelnuts | 138 |
| | Nuts | 128 |
| **Cereals** | Corn | 178 |
| | Rye | 91 – 110 |
| | Barley | 59 – 83 |
| | Wheat | 60 – 69 |
| | Oatmeal | 33 – 52 |
| **Fruits and Vegetables** | Passion fruit | 44 |
| | Orange | 24 |
| | Banana | 14 |
| | Apple | 12 |
| | Pear | 8 |
| | Brussels Sprouts | 43 |
| | Broccoli | 39 |
| | Carrot | 16 |
| | Tomato | 7 |

[a]Data taken from Abidi (2001); Ramos and Saraiva (2009).

According to Phillips and colleagues (2005), nuts and seeds have been constantly associated with the reduction of cholesterol levels and cardiovascular disease incidence in epidemiological studies and clinical trials. They are also rich sources of phytosterols.

Plant stanols can also be found in certain foods, but at much lower concentrations, including some cereals (barley, maize, wheat) and non-hydrogenated oils (Piironen et al., 2000; Trautwein and Demonty, 2007).

Vegetarians typically consume 1 g/day of phytosterols, while other diets can contain lesser amounts (Lagarda et al., 2006). The Western diet contributes a daily intake of 150-400 mg per person (EFSA 2008; Ostlund, 2002; Lagarda et al., 2006; Trautwein and Demonty, 2007), with an average consumption of 65% β-sitosterol, 30% campesterol and 5% stigmasterol (Trautwein et al., 2003). For stanols, a daily average intake is 10% of ingested plant sterols, which is due to its reduced availability in nature (Ramos and Saraiva, 2009).

However, the estimated average intake naturally occurring in the diet is only approximately 10-20% of the recommended intake of 2 g/day to reduce hypercholesterolemia (Saraiva et al., 2011a; Shanahan and Lorimier, 2013.). Through the analysis of Table 1, for a "therapeutic" consumption of 2 g of phytosterols, a substantial amount of maintenance is required, i.e., 4.6 kg of Brussels Sprout, 1.4 kg of sunflower seeds or 621 g of nuts. Therefore, the enrichment of certain food and nutraceutical formulation with phytosterols is of major importance (Mannarino et al., 2014).

## 2.3. Phytosterol Properties

### *2.3.1. Beneficial Effects*

Phytosterols are useful in the pharmaceutical industry because they are used in the production of hormone drugs and also in the cosmetic industry as emulsifiers (Abidi, 2001; Cherif, 2012).

The most important benefit of phytosterols is the reduction of LDL in blood (Trautwein and Demonty, 2007; Mannarino et al., 2014). This effect was established with a daily intake of 2 g of phytosterols by the National Cholesterol Education Program (NCEP) in 2002, which later came to be confirmed by a meta-analysis of Katan and colleagues (2003). Clinical data suggest that daily consumption of 1-3 g phytosterols lowers LDL-cholesterol by 5-15%, without significant effects on HDL cholesterol or triglycerides (Chen et al., 2014; Barbagallo et al., 2015).

In a study conducted by Acuff and co-authors (2007), it was observed that the intake of phytosterol capsules (1.3 g/day) was effective in improving the lipid profile in hypercholesterolemic individuals because, after 3 weeks of treatment, there was a 7% reduction in LDL-cholesterol and a 9% increase in HDL-cholesterol.

Maki and colleagues (2013) also found that the intake of 1.8 g/day of nutraceuticals with phytosterols reduced LDL-cholesterol by 9.2% and total cholesterol by 7.4% and was effective in reducing triglycerides by approximately 9.1% in patients with primary hypercholesterolemia. However, Ottestad and co-authors (2013) also investigated the effect of daily intake of 2 g/day of phytosterols for 4 weeks on plasma levels of LDL-cholesterol and verified that intake of the capsules was not an efficient strategy for patients with mild to moderate hypercholesterolemia because there was only a 2.7% reduction in plasma LDL-cholesterol.

Phytosterols have shown an additive effect on treatment with statins (Ottestad et al., 2013). Goldberg and colleagues (2006) found that tablets with stanols (1.8 g/day) and lecithin produced an additional 9.1% reduction in LDL-cholesterol levels in patients who were already being treated with long-term statins. Thus, stanol/lecithin tablets can be an alternative for individuals who prefer pharmaceutical dosage consistency or who do not consume adequate quantities of foods enriched with phytosterols (Goldberg et al., 2006).

Another study indicating the cholesterol lowering effect of the compounds under study, conducted by Awaisheh and coworkers (2013), assessed the effect of probiotics and phytosterol supplements, either individually or in combination, on serum and liver lipidic profiles in hypercholesterolemic rats. The study found that these supplements were beneficial in reducing serum and liver cholesterol and triglycerides, suggesting the possibility of reducing the risk of cardiovascular diseases.

According to Badimon and co-authors (2010), the nature of food items to which phytosterols are added can substantially affect the reduction of LDL-cholesterol. For instance, when phytosterols were added to milk (15.9%) and yogurt (8.6%), there were significant reductions in LDL-cholesterol, which were considerably lower than those when phytosterols were added to bread (6.5%) and cereals (5.4%).

The cholesterol lowering effect of phytosterols and their safety are well demonstrated scientifically in humans and in animal models (He et al., 2013). However, the mechanism involved has not yet been well established, with various mechanisms proposed for this effect, as will be described in section 2.4. of this chapter.

Beyond the effect of LDL-cholesterol reduction, phytosterols have other beneficial effects, such as the following:

- Anti-inflammatory effects: studies show that dosages from 60 mg/day of β-sitosterol in combination with insignificant amounts of sterolins improve the immune function of individuals affected by various diseases such as pulmonary tuberculosis, human immunodeficiency virus (HIV), allergic reactions and rheumatoid arthritis (Berger et al., 2004; Trautwein and Demonty, 2007; Li et al., 2007; Bernal et al., 2011;. He et al., 2013);
- Antiatherogenic effects: many studies have shown the effect of phytosterols in experimental atherosclerosis animal models with a reduction in atheromas and, as such, a reduction in the development of atherosclerosis (Trautwein and Demonty, 2007);
- Antioxidant activity: according to Berger and colleagues (2004), stanol ester intake can protect LDL cholesterol particles against oxidation. Thus, based on *in vitro* and human studies, there is a possibility that phytosterols may possess antioxidant properties (Berger et al., 2004; Li et al., 2007);
- Anticarcinogenic effects: a study conducted by Awad and Fink (2000) observed anti-tumor effects of phytosterols, particularly β-sitosterol, in cancer cells originating from prostate, colon and breast cancer. Phytosterols have been associated with the inhibition of cancer cell growth, angiogenesis and cancer cell apoptosis in lung, stomach, prostate, ovarian and breast cancer (Berger et al., 2004; Bernal et al., 2011; Sosinska et al., 2013).

### *2.3.2. Adverse Effects*

Phytosterols are recommended by many organizations as a therapeutic way to lower cholesterol, with a daily intake of 2 g. If consumed at the recommended dose, they are well tolerated, with no significant adverse effects, as recognized by the EFSA (2008), so they are classified as a product "generally recognized as safe" (GRAS) by the FDA (Food and Drugs Administration) of the United States of America (USA).

Recent advances in technology have made it possible to combine dietary phytosterols with a variety of foods such as margarines, yogurt, milk, juices, salad, cereal bars and food supplements in capsule or tablet form (Chen et al., 2014), but this situation can lead to the consumption of intake levels higher than the recommended values, which may consequently have an adverse effect.

The combined use of these new products can lead to a cumulative consumption of very high concentrations of phytosterols, which can affect the absorption of carotenoids and liposoluble vitamins (Weingartner et al., 2009). Randomized studies have shown that phytosterols lower blood concentration of β-carotene by approximately 25%, α-carotene by approximately 10% and vitamin E by approximately 8% (Trautwein and Demonty, 2007; Jones and AbuMweis, 2009). Thus, the Scientific Committee on Food (SCF) of the European Union (EU) recommends the consumption of vegetables rich in carotenoids and fruits (EFSA, 2008; Chen et al., 2014).

Another negative effect of phytosterols is phytosterolemia or sitosterolemia, which is a rare hereditary autosomal recessive disease resulting from mutations in the ABCG5 and ABCG8 transporters (Patel, 2008; Myrie et al., 2012). Individuals with sitosterolemia exhibit much higher than normal amounts of phytosterols because they absorb 25-60% of 2 g/day and only a small percentage is excreted through the bile acids (Eussen et al., 2010). For such patients, consumption of phytosterols is contraindicated for having an increased risk of cardiovascular and coronary events and premature atherosclerosis (Katan et al., 2003; IFST, 2011; Myrie et al., 2012; AbuMweis et al., 2014).

Therefore, consumers should avoid excessive or inappropriate intake of products enriched with phytosterols, and the manufacturers of these products should be monitored so they do not adversely affect human health.

## 2.4. Metabolism and Mechanism

Humans absorb 55% to 60% of ingested cholesterol, while the intestinal absorption of plant sterols is very low (Katan et al., 2003; Lagarda et al., 2006; Trautwein and Demonty, 2007; Ellegard et al., 2007; Gonzalez-Larena et al., 2011; Christie, 2012). Studies on absorption, distribution, metabolism and excretion have shown that plant sterols are poorly absorbed from the intestine and the amount that is absorbed is rapidly eliminated via the bile (Table 2).

Although the exact mechanism by which phytosterols induce lower cholesterol levels in blood remains to be clarified, several hypotheses have been developed. In the 1960s, it was believed that phytosterols compete with cholesterol for incorporation into mixed micelles. Ever since, it has been suggested that phytosterols interfere with the incorporation of cholesterol in chylomicron (Woodgate et al., 2006; Shaghaghi et al., 2013; Chen et al., 2014), and for this reason, it is assumed that they regulate various genes involved in cholesterol homeostasis (He et al., 2013; Barbagallo et al., 2015). Moreover, phytosterols stimulate expression of ATP transporter - binding cassette - because these transporters, especially ABC1, do not distinguish cholesterol from phytosterols, which can promote the efflux of phytosterols into the intestinal lumen (Trautwein and Demonty, 2007; Eussen et al., 2010; Mannarino et al., 2014).

**Table 2. Average intake levels, sources, and metabolism of dietary sterols[(b)]**

| | Cholesterol | Plant sterols | Plant stanols |
|---|---|---|---|
| **Dietary intake** | 300–500 mg/day | 200–400 mg/day | 10–60 mg/day |
| **Dietary sources** | Egg yolk, butter, dairy products, meat, crustaceans | Vegetable oils, corn (maize), beans, grains | Grains (wheat, rye, corn/maize) |
| **Endogenous synthesis** | 800–1200 mg/day | Not synthesized | Not synthesized |
| **Absorption** | 40 – 60% | 5 – 15% | 0.1 – 2% |
| **Plasma concentration** | 140 – 320 mg/dL | 0.3 – 1.7 mg/dL | 0.01 mg/dL |
| **Excretion** | 40 – 60% | 85 – 95% | >98% |

[(b)]Data taken from Salo et al. (2003).

Other possible effects of phytosterols include reduction of apoB secretion from enterocytes and hepatocytes, reduction of cholesterol synthesis (Eussen et al., 2010), activation of cellular stress responses and modulation of the inflammatory cascade (Mannarino et al., 2014).

However, incorporating plant sterols/stanols within various emulsion based delivery systems to prevent agglomeration in the small intestine, as well as reducing their crystal particle size, can improve the cholesterol-lowering efficacy (Shaghaghi et al., 2013; Maki et al., 2013).

## 3. Nutraceuticals Enriched with Phytosterols

The food industry has focused greatly on food fortification with phytosterols, classifying them as functional foods because they serve the function of improving health and reducing the risk of contracting a disease (Mannarino et al., 2014). The pharmaceutical industry ascertains the focus of these foods and formulates dietary supplements in the form of tablets or capsules, to which phytosterols may be added, which are designated as nutraceuticals.

Nutraceuticals are dietary supplements that provide a concentrated form of a bioactive agent from a food within a non-food matrix. They are used for the purpose of improving health in doses exceeding those that may be obtained from normal food (Bernal et al., 2011). Recent studies have shown promising results in various pathological complications such as diabetes, atherosclerosis, cardiovascular diseases, cancer and neurological disorders (Barbagallo et al., 2015).

The increasing interest in and use of nutraceuticals for the population has made the dietary supplement industry one of the most lucrative healthcare markets. It is anticipated to reach $ 250 billion by 2018 (Das et al., 2012).

## 3.1. Functional Foods *vs* Nutraceuticals: Effectiveness

The production of nutraceuticals or functional foods intends to use food components to improve or cure certain diseases or help maintain good health.

Nutraceuticals are often used without significant evidence, i.e., action mechanisms have not been clearly confirmed, most clinical data are derived from small uncontrolled studies and few clinical trials document the relationship between these interventions.

Currently in the EU, in addition to margarines, which were the first functional foods with phytosterols, numerous functional foods have been approved, including yogurt, milk, cheese spreads, soy, cereal bars, fruit juices, breads and sausages (Weingartner et al., 2009; Maki et al., 2013; Barbagallo et al., 2015). However, nutraceuticals with phytosterols can be a more practical and effective method to aid hypercholesterolemic patients who are not consuming the previously mentioned foods every day (Woodgate et al., 2006; Sorrentino, 2008; AbuMweis et al., 2014).

In fact, the tablets or capsules for phytosterols have been suggested to be more flexible and convenient for continuous doses. Furthermore, they are more accessible, with other pharmaceutical/nutraceutical cholesterol lowering agents to enhance their therapeutic and preventive efficacy as adjuvant therapy with other tablets (McPherson et al., 2005; Shaghaghi et al., 2013).

Barbagallo and colleagues (2015) reported that individuals with a medium cardiovascular risk may benefit from treatment with nutraceuticals, where a moderate reduction in LDL-cholesterol may be sufficient to achieve internationally recommended levels (≤190 mg/dl). Nutraceuticals can also be useful when statin therapy in normal dosage is not effective. The same authors mention that nutraceuticals can be applied to patients who are afraid of the side effects of conventional drugs or simply do not want to be considered a patient.

The various dietary guidelines that recommend phytosterol intake to reduce high LDL-cholesterol levels do not distinguish between the intake of margarine, tablets or other matrices (Jones and AbuMweis, 2009; Ottestad et al., 2013). The available scientific evidence indicates that the types of vehicle to which phytosterols are added do not affect their effectiveness, but few studies have administered phytosterols in the pharmaceutical form (Woodgate et al., 2006; FDA, 2010; Maki et al., 2013; Ottestad et al., 2013).

With regard to the additive effect of some nutraceuticals, combination therapy with classical drugs can improve clinical outcomes, which leads us to consider that nutraceuticals may be a hypolipidemic alternative treatment and, if properly used, can represent a valid strategy for cardiovascular disease prevention.

## 3.2. Nutraceutical Safety Use

There is a wide variety of nutraceuticals with therapeutic effects that are easily accessible to patients. Because they are considered "food," their consumption is said to be safe, with no significant toxicity (Bernal et al., 2011). However, the demand has increased for nutraceuticals, and they are, in most cases, used as self-medication to compensate for certain food shortcomings or to control pathological situations, which can lead to an indiscriminate and excessive use that may increase the risk of developing adverse reactions or even toxicity.

Their use does not follow the same rules of classic medicines. The process of drug marketing is very lengthy, starting with *in vitro* testing for possible effects, followed by evaluation in animal models and in humans and finally analyses of effectiveness and tolerability. After approval and availability in the market, there is also a close monitoring of side effects and ultimately controlled clinical trials to validate the results. In contrast, nutraceuticals are used as therapy without significant evidence (Barbagallo et al., 2015).

According to the report of the Conference of the National Institutes of Health (NIH, 2006) on nutraceuticals in preventing chronic diseases, it is important to carry out further studies on the efficacy and safety of nutraceuticals to reach firm conclusions or make recommendations for or against these supplements. Beyond that, the possibility of interactions between nutraceuticals and drugs administered concurrently cannot be neglected. Thus, more scientific evidence is required for a better understanding of the main contributions of nutraceuticals in inhibiting the progression of cardiovascular disease (Badimon et al., 2010).

With the increasing number of nutraceuticals and functional foods with phytosterols, it is important to implement measures that protect the health of consumers. Such measures are analyzed in the following section.

## 3.3. Legislation

There is no specific legislation in Europe to control nutraceuticals, and they therefore end up being controlled as food supplements or drugs depending on the manufacturer and product positioning in the market (Bernal et al., 2011). In the EU, they are regulated by Directive 2002/46/EC focusing on vitamins and minerals that are used as ingredients of food supplements (EFSA, 2013). The directive sets labeling requirements and establishes the maximum and minimum levels for each vitamin and mineral added to supplements. In addition, Annex II, which was amended by the EU 1170/2009, contains a list of permitted vitamin or mineral substances that may be added for specific nutritional purposes in food supplements (EC, 2002 and 2009). The nutraceuticals may also contain other substances with a nutritional or physiological effect that are not included in the allowed list if the European Commission (EC) authorizes such inclusion (EFSA, 2013).

In the USA, the FDA regulates nutraceuticals under a different set of guidelines than those that control food and medicine (Bernal et al., 2011), ensuring the safety, wholesomeness and labeling to be truthful and not misleading (FDA, 2015). Similarly, the EC regulates the nutraceutical market through the EFSA, which authorizes the labeling of food products with health claims, particularly the food and nutraceuticals enriched with phytosterols with the following words: "The plant sterols reduce cholesterol in the blood. The reduction in blood cholesterol can reduce the risk of coronary artery disease" (EC, 2006; EFSA, 2008).

Companies intending to use a health claim provide information to a national authority of a Member State of the EFSA, which subsequently issues an opinion on the health claim supported by scientific evidence and whether it is in accordance with EU 1924/2006 (EC, 2006; Hobbs et al., 2014). The EU regulatory system has a relatively strict approach compared to the USA with more applications rejected than accepted (Mannarino et al., 2014). In contrast, some authors characterize the USA regulatory system as relaxed, as it does not require prior approval from the FDA to market nutraceuticals (Hobbs et al., 2014; FDA, 2015).

In respect to phytosterols, their application is controlled by EC Regulation no.258/97, which relates to foods and food ingredients. For a novel food or food ingredient to be placed on the market, it needs a specific authorization, involving a rigorous safety assessment process. As previously mentioned, phytosterols have been classified as a safe ingredient by the FDA and were approved in Europe, by the SCF, to be used in margarines. According to the Committee of Experts of the FAO/WHO on Food Additives (JECFA), in 2008, various approvals were granted to different types of food. Importantly, products not fortified with stanols need not comply with this Regulation, as they were on the market as food before the introduction of this legislation.

However, it is clear that manufacturers must follow the recommended daily intake according to the EFSA (2008) of 1.5-3 g for an adult with medium size characteristics or according to the NCEP (2002), which recommends 2 g daily intake of phytosterols to lower elevated levels of LDL cholesterol.

Following the SCF's recommendations, the EC adopted specific regulations for the labeling of foods and nutraceuticals enriched with phytosterols and their respective esters (EC, 2004a). Thus, every label must contain in visible and legible words the sentence," with added plant sterols/plant stanols;" the amount of added phytosterols, phytosterol esters, phytostanols or phytostanol ester content (expressed in % or as g of free plant sterols/plant stanols per 100 g or 100 ml of the food) in the list of ingredients; a statement that the product is exclusively for people who want to lower their blood cholesterol levels, not exceeding a consumption of more than 3 g/day of added phytosterols; an indication that patients with medication to lower cholesterol level should only consume the product under medical supervision; an indication, easily visible and legible, stating that the product may not be nutritionally appropriate for pregnant or breastfeeding women and children under the age of five years; a recommendation for the integration of the product of consumption with a balanced and varied diet, including regular consumption of fruit and vegetables to help maintain carotenoid levels; and a setting portion of the food or food ingredient concerned (preferably in g or ml), with a statement of the amount of plant sterols/plant stanols contained in each dose.

In addition to all these European standards, products enriched with phytosterols to be placed on the market must conform to a profile of sterols and plant stanols, according to regulations 2004/333/EC, 2004/334/EC, 2004/335/EC and 2004/336/EC, where a mixture that contains 99% of naturally occurring phytosterols would be formed by β-sitosterol (<80%), β-sitostanol (<15%), campesterol (<40%), campestanol (<5%), stigmasterol (<30%), brassicasterol (<3%) and other (<3%) (CE, 2004b, c, d and e).

Nonetheless, legislation for nutraceuticals can in some cases not be sufficient, and for this reason, Europe has joined with associations and companies of such products to ensure that future legislation and community policies reflect the importance of the role this sector plays in the health of consumers.

## 3.4. Analytical Methodologies for the Determination of Phytosterols in Nutraceuticals

Phytosterol analysis in nutraceuticals is essential for analytical quality assessment and evaluation of possible adulteration of the products marketed. In this regard, it is highly necessary to develop and apply analytical techniques (Bernal et al., 2011). The analytical determination of phytosterols is often a difficult task due to the polarity of these substances and the low content and availability in food matrices.

Initially, phytosterols were determined by enzymatic and spectrophotometric methods; however, problems with interference and shortness of specificity in the determination of the total amount of phytosterol has led to the adoption of chromatographic methods (Ramos and Saraiva, 2009).

There are no standard methods developed for analysis of nutraceutical phytosterols. Due to the complexity of such natural matrices, the use of analytical techniques (such as mass spectrometry (MS), Nuclear Magnetic Resonance (NMR), high performance liquid chromatography (HPLC), Capillary Electrophoresis (CE), NMR-HPLC, HPLC-MS, GC-MS and CE-MS) is indispensable (Bernal et al., 2011).

Determination of phytosterols is usually carried out by gas chromatography (GC) but can also be performed by high performance liquid chromatography (HPLC) (Lagarda et al., 2006) and by infrared Fourier transform spectroscopy (FTIR) (Yan et al., 2011). HPLC has the advantage of working at much lower column temperatures, which may be appropriate for analyzing this type of thermally unstable compound. However, the high lipophilicity of phytosterols can make sample preparation and its chromatographic analysis difficult (Abidi, 2001; Lagarda et al., 2006). This methodology cannot provide sufficient selectivity in the separation of sterols and stanols (Ramos and Saraiva, 2009), and thus, there may be a risk of possible interference from compounds other than those of interest (Ahmida et al., 2006). Thus, GC coupled with flame ionization (GC-FID) and especially mass spectrometry (GC-MS) is the best and most used option for the separation, identification and quantification of phytosterols (Ahmida et al., 2006; FDA, 2010; Saraiva et al., 2011b; Winkler-Moser, 2011).

In this part of the chapter, only the GC-MS methodology developed for the identification and quantification of phytosterols in nutraceuticals is addressed.

### *3.4.1. Sample Preparation*

The first and most important step in the analysis is phytosterol sample preparation. The objectives at this stage are to isolate the fraction of sterol esters and conjugates by converting all phytosterols into free or esterified phytosterols to be analyzed by chromatography. This process depends on the matrix and the original form of the phytosterol in the sample (Lagarda et al., 2006; Winkler-Moser, 2011).

The research was based on a nutraceutical with the addition of phytosterols, specially formulated for adults who want to maintain healthy cholesterol levels in the blood. This nutraceutical is presented to the consumer in the form of tablets, with a recommended consumption of 2 tablets once a day (2 x 500 mg phytosterols), during or after meals.

### *3.4.2. Lipid Extraction*

Phytosterols present in plants or in vegetable oils can be extracted by solvents containing chloroform-methanol, chloroform-methanol-water, *n*-hexane, dichloromethane or acetone

(Lagarda et al., 2006). The largest problem for extraction with non-polar solvents is that they may extract other unsaponifiable substances that will interfere with the detection and quantification of substances in trace concentrations.

According to Largarda and colleagues (2006), when comparing the extraction with the solvent with extraction by SPE (Solid Phase Extraction), extraction with the solvent provides a faster fractionation with lower losses of analytes and uses smaller volumes of solvent.

### *3.4.3. Saponification*

Saponification is a crucial step for the conversion of phytosterols to a form that is soluble in nonpolar solvents. According to Lagarda and co-authors (2006), acid hydrolysis with hydrochloric acid, alkaline saponification with potassium hydroxide and extraction with toluene have been used to determine the phytosterols in cereal products. However, saponification charges are quite complex for routine analysis, and acid hydrolysis is not applicable to all phytosterol structures, in particular $\Delta_7$-phytosterols $\Delta_7$-avenasterol and $\Delta_7$-stigmastanol due to decomposition or isomerization after a short period of acid hydrolysis. Alkaline saponification with ethanolic potassium hydroxide with the addition of an internal standard, either at room temperature or by heating, followed by extraction of the unsaponifiable material has been adequate for most matrices enriched with phytosterols, including nutraceuticals (Lagarda et al., 2006; Ahmida et al., 2006; Santos et al., 2007; FAO, 2008).

### *3.4.4. Extraction of the Unsaponifiable Fraction*

After saponification, a solvent or mixture of solvents for extraction of the substances of interest is added. Generally, in food matrices, biological samples and arrays in the form of tablets, the unsaponifiable fraction of phytosterols is extracted by an apolar solvent, in particular by *n*-hexane (Ahmida et al., 2006; Santos et al., 2007; Saraiva et al., 2011a and b). However, other solvents may be applicable, such as heptane, diethylether, chloroform, or their blends (Lagarda et al., 2006; Winkler-Moser, 2011).

### *3.4.5. Determination by GC-MS of Phytosterols in Nutraceuticals*

In GC analysis, volatility is an essential requirement because the phytosterols have low volatility and are relatively labile substances at elevated temperatures, whereby the chemical conversion to trimethylsilyl ethers or acetylated derivatives is an indispensable step (Abidi, 2001; Lagarda et al., 2006). However, trimethylsilyl derivatives are simpler and form more rapidly, in addition to being more volatile substances and analogously less polar, with higher molecular weight and overall better chromatographic behavior than the acetylated derivatives (Abidi, 2001).

The derivatization reagent most used in phytosterol analysis, according to Lagarda and colleagues (2006), is N, O-*bis*-trimethylsilyltrifluoroacetamide (BSTFA) with 1% trimethylchlorosilane (TMCS) and N-methyl-N-trimethyl silyltrifluoroacetamide (MSTFA) dissolved in pyridine. However, Saraiva and authors (2011b) showed that the mixture of N-methyl-N-trimethylsilyl-trifluoroacetamide (MSTFA)/1,4-dithioerythritol (DTE)/tri-methyliodosilane (TMIS) also gives good results when added to the residue of the sample, previously dried in an atmosphere of nitrogen.

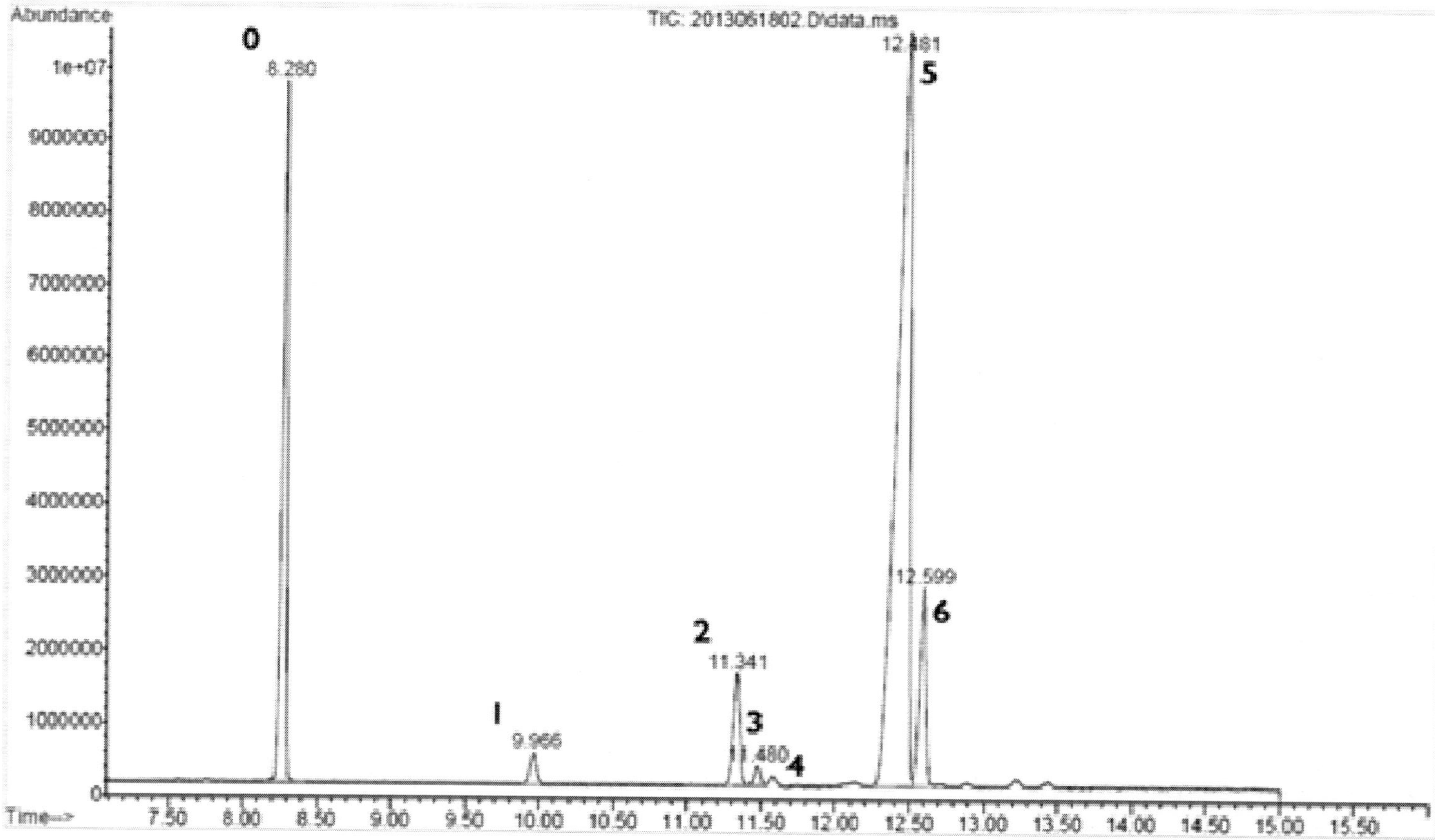


Figure 1. Chromatogram of a nutraceutical enriched with phytosterols. See peak legend in Table 3.

**Table 3. Retention times (RT) and ion transitions for identification of the phytosterols present in a nutraceutical**

| | | | | Major fragmentation ions *m/z* | | | | |
|---|---|---|---|---|---|---|---|---|
| Peak | Compounds | RT | $[M]^+$ | $[M\text{-}15]^+$ | $[M\text{-}90]^-$ | $[M\text{-}105]^-$ | $[M\text{-}129]^-$ | Others |
| 0 | cholestane(IS) | 8.280 | 372 | 357 | | | | 217(1) |
| 1 | α-tocopherol(2) | 9.966 | 502(1) | 487 | | | | 237;277 |
| 2 | campesterol | 11.341 | 472 | 457 | 382 | 367 | 343 | 129;145;255;343(1) |
| 3 | campestanol | 11.480 | 474 | 459 | 384 | 369 | 345 | 215(1);305;417 |
| 4 | stigmasterol | 11.579 | 484 | 469 | 394 | 379 | 355 | 129(1);255 |
| 5 | β-sitosterol | 12.481 | 486 | 471 | 396 | 381 | 357(1) | 129;255 |
| 6 | β-Sitostanol | 12.599 | 488 | 473 | 398 | 383 | 359 | 215(1); 305; 431 |

(1) Most abundant ions (diagnostic ion).

(2) Mottier et al. (2002 and 2006).

In this type of methodology, the addition of an internal standard (IS) to the sample is needed to compensate for losses during the analytical process and therefore obtain a correct quantification of the compounds (Abidi, 2001). The internal standard is added in a known concentration to all samples to be analyzed, which should preferably be placed before the extraction process. However, it may also be added after extraction but prior to derivatization.

Considering the intrinsic characteristics of the substances to be analyzed, the IS selected for the quantification of phytosterols in nutraceuticals is cholestane because it is unlikely to interfere with other sterols and presents a very similar response to phytosterols. However, other internal standards may be added, such as betulin, coprostanol and cholestanol (Lagarda et al., 2006).

As mentioned earlier, the GC-MS technique is more appropriate to the identification of phytosterols because it allows analysis by selective ion monitoring (SIM) or a full analysis of all ions (SCAN) and enables the assessment of the purity of the eluted peak (Abidi, 2001). The chromatographic process of phytosterols in nutraceutical analysis was based on the method proposed by Santos and colleagues (2007) and Saraiva and co-authors (2011b), allowing the identification of five phytosterols and a vitamin (Figure 1). Table 3 shows the retention times (RT) and interpretation of the fragmentation of unsaponifiable compounds in different ions of nutraceuticals determined by GC-MS.

Identification of phytosterols was achieved by comparing the retention times of the chromatographic peaks obtained in the sample with those of standards and their respective mass spectra. Quantification of the phytosterols present in the nutraceutical was determined by comparing the ratios obtained from the peak areas of the sample phytosterols with the internal standard of each chromatogram.

Using the analytical methodology developed for the identification and quantification of phytosterols in nutraceuticals, it was found that β-sitosterol (298.26 mg $\mu g^{-1}$) was the most abundant substance, followed by β-sitostanol (133.29 mg $\mu g^{-1}$), campesterol (49.87 mg $\mu g^{-1}$), campestanol (19.76 mg $\mu g^{-1}$) and stigmasterol (7.71 mg $\mu g^{-1}$).

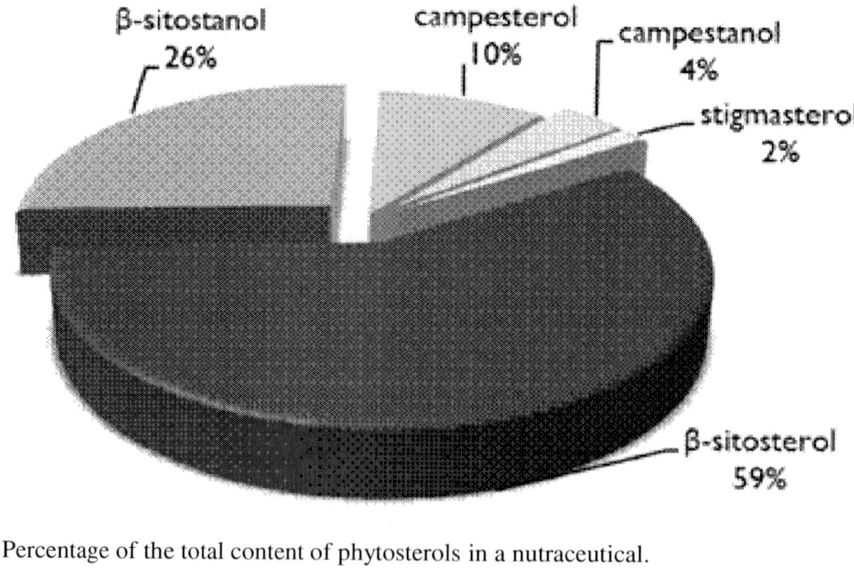


Figure 2. Percentage of the total content of phytosterols in a nutraceutical.

Analyzing the profile of sterols and stanols in nutraceuticals showed that a certain percentage composition is not in accordance with EU specifications for functional foods (CE, 2004b, c, d and e). Because the percentage of β-sitostanol content (Figure 2) is higher than 15%, it is not in agreement with the current legislation. Nonetheless, this fact may not be relevant because there are marketed products that only include stanols, and there is no scientific evidence of harmful health effects from higher doses.

#### 3.4.5.1. Quality Control in Nutraceuticals

Quality control and standardization of nutraceuticals, the determination of the real bioavailability of different compounds, and the assessment of the effects of these bioactive compounds on health constitute a huge challenge. Together with the questions on safety and toxicity, product stability and the possibility of adulteration, developing methodologies to better understand the different compounds used for the preparation of nutraceuticals becomes important.

The nutraceutical manufacturers suggest the use of two tablets daily, equivalent to 1 g/day of phytosterols. However, with the analytical technique developed, the concentration found (1.4 g/day) could be damaging if consumed with other functional foods, nutraceuticals or hypocholesterolemic agents. However, if consumed with a balanced diet and without the addition of other dietary supplements, it can be considered safe because it does not exceed the recommended daily intake of phytosterols.

Nutraceuticals can also be effective in reducing LDL-cholesterol, as indicated on the label, if they contain sterols and plant stanols in significant quantities. However, to verify the effectiveness of the supplement, a clinical trial in hypercholesterolemic individuals should be conducted, and in fact, more studies are needed to understand whether the recommended daily dose of phytosterols is affected when contained within other matrices.

## CONCLUSION

Scientific evidence has shown that the quality control of food products marketed for the prevention of cardiovascular diseases has been crucial for preventing the possible adulteration of these products posing a risk to consumer's health. It is therefore essential to ensure the safety of food products enriched with phytosterols through innovative procedures to meet market requirements and consumer preferences.

This chapter intends to contribute directly to the improvement of the quality control of such products as well as to food safety for consumers, an important area between food and health.

Nutraceuticals represent a valid alternative in the hypolipidemic treatment and have therefore had an essential role in cardiovascular prevention strategies. With the growing industrial and commercial interest, it is important to have more explicit regulations and to develop scientific research in this area.

## REFERENCES


Abidi S.L. (2001) – Review: Chromatographic analysis of plant sterols in foods and vegetable oils. *Journal of Chromatography A*, 935, pp. 173–201.

AbuMweis, S.S., Marinangeli, C.P.F., Frohlich, J., Jones, P.J. (2014). Implementing Phytosterols into Medical Practice as a Cholesterol-Lowering Strategy: Overview of Efficacy, Effectiveness and Safety. *Canadian Journal of Cardiology*, doi: 10.1016/ j.cjca.2014.04.022.

Acuff R.V., Cai D. J., Dong Z-P., Doris B. (2007) - The lipid lowering effect of plant sterol ester capsules in hypercholesterolemic subjects. *Lipids in Health and Disease*, 6 – 11.

Ahmida H.S.M., Bertucci P.; Franzò L., Massoud R., Cortese C., Lala, A., Frederici G. (2006) - Simultaneous determination of plasmatic phytosterols and cholesterol precursors using gas chromatography –mass spectrometry (GC–MS) with selective ion monitoring (SIM). *Journal of Chromatography B*, 842, pp. 43–47.

Awad A., Fink C. (2000) – Phytosterols as Anticancer Dietary Components: Evidence and Mechanism of Action. *American Society for Nutritional Sciences*. 130, 2127 – 2130 [Cited 2012.07.07]. Available from: http://jn.nutrition

Awaisheh S.S., Khalifeh M.S., Al-Ruwaili M.A., Khalil O.M., Al-Ameri O.H., Al-Groom R. (2013) - Effect of supplementation of probiotics and phytosterols alone or in combination on serum and hepatic lipid profiles and thyroid hormones of hypercholesterolemic rats. *Journal of Dairy Science*, 96, 9 – 15.

Badimon, L., Vilahur, G. and Padro, T. (2010). Nutraceuticals and Atherosclerosis: Human Trials. *Cardiovascular Therapeutics*, 28, 202 – 215.

Barbagallo, C. M., Cefalu, A.B., Noto, D. and Averna, M. (2015). Role of Nutraceuticals in Hypolipidemic Therapy. *Frontiers in Cardiovascular Medicine*, 2:22. doi:10.3389/fcvm. 2015.00022

Berger A., Jones P. JH., Abumweis S.S. (2004) - Plant sterols: factors affecting their efficacy and safety as functional food ingredients. *Lipids in Health and Disease*, 3:5. [Cited 2012.11.04]. Available from: http://lipidworld.com/content/3/1/5

Bernal, J., Mendiola, J.A., Ibá˜nez, Cifuentes, E.A. (2011). Review: Advanced analysis of nutraceuticals. *Journal of Pharmaceutical and Biomedical Analysis*, 55, 758 – 774.

Chen, G., Wang, H., Zhang, X. and Yang, S. T. (2014). Nutraceuticals and Functional Foods in the Management of Hyperlipidemia. *Critical Reviews in Food Science and Nutrition,* 54:9, 1180-1201, doi: 10.1080/10408398. 2011.629354

Cherif A. A. (2012) - Phytochemicals Components as Bioactive Foods, Bioactive Compounds in Phytomedicine, Prof. Iraj Rasooli (Ed.) *InTech.* ISBN: 978-953-307-805-2.

Christie, W.W. (2012) - Sterols and their conjugates from plants and lower organisms: structure, occurrence, biochemistry and analysis. *The Lipid Library*. Sterols 3. [Cited 2012.06.28]. Available from: http://lipidlibrary. aocs.org/

Clifton P.M., Noakes M., Sullivan D., Erichsen N., Ross D., Annison G., Fassoulaki A., Cehun M., Nestel P. (2004) - Cholesterol-lowering effects of plant sterol esters differ in milk, yoghurt, bread and cereal. *European Journal of Clinical Nutrition*, 58, 503 – 509.

Das, L., Bhaumik, E., Raychaudhuri, U., Chakraborty, R. (2012). Role of nutraceuticals in human health. *Journal Food Science Technology*, 49, 173 – 83.

EC (1997). Regulation (EC) No 258/97 of the European Parliament and of the Council of 27 January 1997 concerning novel foods and novel food ingredients. *Official Journal of the European Union*, L043, 1-7.

EC (2002). Commission Decision 2002/46/EC of the European Parliament and of the Council of 10 June 2002 on the approximation of the laws of the Member States relating to food supplements. *Official Journal of the European Communities,* L183, 51-57.

EC (2004a). Commission Decision 2004/608/EC of 31 March 2004 concerning the labeling of foods and food ingredients with added phytosterols, phytosterol esters, phytostanols and/or phytostanol esters. *Official Journal of the European Union*, L97, 44-45.

EC (2004b). Commission Decision 2004/333/EC of 31 March 2004 authorising the placing on the market of yellow fat spreads, salad dressings, milk type products, fermented milk type products, soya drinks and cheese type products with added phytosterols/phytostanols as novel foods or novel food ingredients under Regulation (EC) No 258/97 of the European Parliament and of the Council. *Official Journal of the European Union*, L105, 40-42.

EC (2004c). Commission Decision 2004/334/EC of 31 March 2004 authorising the placing on the market of yellow fat spreads, milk type products, yoghurt type products, and spicy sauces with added phytosterols/phytostanols as novel foods or novel food ingredients under Regulation (EC) No 258/97 of the European Parliament and of the Council. *Official Journal of the European Union*, L105, 43-45.

EC (2004d). Commission Decision 2004/335/EC of 31 March 2004 authorising the placing on the market of milk type products and yoghurt type products with added phytosterol esters as novel food ingredients under Regulation (EC) No 258/97 of the European Parliament and of the Council. *Official Journal of the European Union*, L105, 46-48.

EC (2004e). Commission Decision No 2004/336/EC of 31 March 2004 authorising the placing on the market of yellow fat spreads, milk based fruit drinks, yoghurt type products and cheese type products with added phytosterols/phytostanols as novel foods or novel food ingredients under Regulation (EC) No 258/97 of the European Parliament and of the Council. *Official Journal of the European Union*, L105, 49-51.

EC (2006). Commission Regulation No 1924/2006 of the European Parliament and of the Council of 20 December 2006 on nutrition and health claims made on foods. *Official Journal of the European Union*. L404, 9-25.

EC (2009). Commission Regulation No 1170/2009 of 30 November 2009 amending Directive 2002/46/EC of the European Parliament and of Council and Regulation (EC) No 1925/2006 of the European Parliament and of the Council as regards the lists of vitamin and minerals and their forms that can be added to foods, including food supplements. *Official Journal of the European Union*. L314, 36-42.

EFSA (2008) - Consumption of Food and Beverages with Added Plant Sterols in the European Union. *The EFSA Journal*, 133, 1-21.

EFSA (2012) - Opinion of the Scientific Panel on Food additives and Nutrient Sources added to Food (ANS) on a request from the Commission on the safety of stigmasterol-rich plant sterols as food additive. *EFSA Journal*, 10, 1 – 39.

EFSA (2013). Food supplements: European Food Safety Authority. [Cited 2015.03.05]. Available from: http://www.efsa.europa.eu/en/topics/topic /supplements.htm

Ellegard L.H, Andersson S.W., Normén A.L., Andersson H.A (2007) - Dietary Plant Sterols and Cholesterol Metabolism. *Nutrition Reviews*, Vol. 65, N°1, pp. 39 – 45.

Eussen S., Klungel O., Garssen J., Verhagen H., Kranen v. H., Loveren v.H., Rompelberg C. (2010) - Support of drug therapy using functional foods and dietary supplements: focus on statin therapy. *The British Journal of Nutrition*, 103, 9, 1260 – 1277.

FAO (2008) - Phytosterols, phytostanols and their esters: chemical and technical assessment. [Cited 2012.05.28]. Available from: http://www.fao.org/

FDA (2010). Proposed Rule: Food Labeling: Health Claim; Phytosterols and Risk of Coronary Heart Disease, vol. 75, No. 235.

FDA (2015). U.S. Food and Drug administration [internet]. Silver Spring: U.S. Food and Drug Administration; U.S. Department of Health and Human Services. [updated 2015.04.29; cited 2015.05.15]. Available from: http://www.fda.gov/aboutfda /transparency/basics/ucm193949.htm

Goldberg A.C., Ostlund R.E., Bateman J.H., Schimmoeller L., McPherson T. B., Spilburg C.A. (2006) - Effect of Plant Stanol Tablets on Low-Density Lipoprotein Cholesterol Lowering in Patients on Statin Drugs. *The American Journal of Cardiology*, 97, 376 – 379.

Gonzalez-Larena M., García-Llatas G., Vidal M.C., Sánchez-Siles L.M., Barberá R., Lagarda M.J. (2011) - Stability of Plant Sterols in Ingredients Used in Functional Foods. *Journal of Agricultural and Food Chemistry*, 59, 3624 – 3631.

He W.S., Wang M.G., Pan X.X., Li J.J., Jia C.S., Zhang X.M., Feng B. (2013) - Role of plant stanol derivatives in the modulation of cholesterol metabolism and liver gene expression in mice. *Food Chemistry*, Volume 140, 9 – 16.

Hobbs, J. E., Malla, S. and Sogah, E. K. (2014). Regulatory Frameworks for Functional Food and Supplements. *Canadian Journal of Agricultural Economics*, 62, 569–594.

IFST (2011) - Phytosterol Esters (Plant Sterol and Stanol Esters). Institute of Food Science and Technology, Information Statement.

Jones P.J.H., AbuMweis S. S. (2009) - Phytosterols as functional food ingredients: linkages to cardiovascular disease and cancer. Current Opinion in *Clinical Nutrition and Metabolic Care* 12, 147–151. ISSN 1363-1950.

Katan M.B, Grundy S.M., Jones P., Law M., Miettinen T., Paoletti R. (2003) - Efficacy and safety of plant stanols and sterols in the management of blood cholesterol levels. *Mayo Clin Proc.*Vol.78, pp. 965 - 978.

Lagarda M.J, Garcia-Llatas G., Farré R. (2006) - Analysis of phytosterols in foods. *Journal of Pharmaceutical and Biomedical Analysis*, 41, 1486–1496.

Li T.S.C., Beveridge T.H.J., Drover J.C.G. (2007) - Phytosterol content of sea buckthorn (*Hippophae rhamnoides L.*) seed oil: Extraction and identification. *Food Chemistry*, 101, 1633 – 1639.

Lutjohann D. (2004) - Sterol autoxidation: from phytosterols to oxyphytosterols. *Journal of Nutrition*, 91, 3-4.

Maki K.C., Lawless A.L., Reeves, M.S., Kelley K.M., Dicklin M.R., Jenks B.H., Shneyvas E., Brooks J.R. (2013) - Lipid effects of a dietary supplement softgel capsule containing plant sterols/stanols in primary hypercholesterolemia. *Nutrition*, 29, 1, 96 – 100.

Mannarino, M.R., Ministrini, S., Pirro, M. (2014). Review Article: Nutraceuticals for the treatment of hypercholesterolemia. *European Journal of Internal Medicine*, 25, 592–599.

McPherson, T. B., Ostlund, R. E., Goldberg, A. C., Bateman, J. H., Schimmoeller, L. and Spilburg, C. A. (2005). Phytostanol tablets reduce human LDL-cholesterol. *Journal of Pharmacy and Pharmacology*, 57, 889–896.

Mottier, P. and Guy, P. A. (2006). GC–MS and LC–MS/MS Techniques to Quantify α-Tocopherol and α-Tocopherolquinone: Applications to Biological and Food Matrices. *The Encyclopedia of Vitamin E* (eds V.R. Preedy and R.R. Watson), 190-200.

Mottier, P., Gremaud, E., Guy, P. A. and Turesky, R. J. (2002). Comparison of Gas Chromatography-Mass Spectrometry and Liquid Chromatography-Tandem Mass Spectrometry Methods to Quantify α-Tocopherol and α-Tocopherolquinone Levels in Human Plasma. *Analytical Biochemistry*, 301, 128–135.

Myrie, S.B, Mymin, D., Triggs-Raine, B. and Jones, P.J.H. (2012). Serum lipids, plant sterols, and cholesterol kinetic responses to plant sterol supplementation in phytosterolemia heterozygotes and control individuals. *The American Journal of Clinical Nutrition*, 95, 837–44.

NCEP (2002). National Health, Heart, Lung, and Blood Institute, Health. Third Report of the *National Cholesterol Education Program* (NCEP) Expert Panel on Detection, Evaluation and Treatment of High Blood Cholesterols in Adults (Adult Treatment Panel III) Final Report. *National Institute of Health Publication*. 02 - 5215.

NIH (2006) – Multivitamin/Mineral Supplements and Chronic Disease Prevention. *National Institute of Health*, State-of-the-Science Conference Statement, Vol.23, no.2.

Ostlund, R.E. (2002). Phytosterols in human nutrition. *Annual Review of Nutrition*, 22, 533-549.

Ottestad I., Ose L., Wennersberg M.H., Granlund L., Kirkhus B., Retterstøl K. (2013) - Phytosterol capsules and serum cholesterol in hypercholesterolemia: A randomized controlled trial. *Atherosclerosis*, 228, 421 – 425.

Patel S.B. (2008) - Plant Sterols and Stanols: Their Role in Health and Disease. *Journal of Clinical Lipidology*, 2 (2), S11 – S19.

Phillips K. M., Ruggio D. M., Ashraf-Khorassani M. (2005) - Phytosterol composition of nuts and seeds commonly consumed in the United States. *Journal Agriculture Food Chemistry*, 53, 9436 – 9445.

Piironen V., Toivo J.; Lampi A.M. (2000): Natural sources of dietary plant sterols. *Journal of Food Composition and Analysis*, 13, 619-624.

Ramos F., Saraiva D. (2009) - Milk consumption and health. *Nova Science*, New York. ISBN: 978- 1-60741-459-9.

Salo, P., Wester, I., Hopia, A., Gunstone, F.D. (2003). Phytosterols. In: Lipids for functional foods and nutraceuticals. Ed. Anonymous. Bridgwater, England: *The Oily Press*, Vol.13, 183-224.

Santos R., Limas E., Sousa M., Castilho M.C., Ramos F., Noronha da Silveira M.I. (2007) - Optimization of analytical procedures for GC–MS determination of phytosterols and phytostanols in enriched milk and yoghurt. *Food Chemistry*, 102, 113 – 117.

Saraiva D., Castilho M.C., Martins M.R., Noronha da Silveira M.I., Ramos F. (2011a) - Evaluation of phytosterols in milk and yogurts used as functional foods in Portugal. *Food Anal. Methods*, 4, 28 – 34.

Saraiva D., Semedo R., Castilho M.C, Silva J.M, Ramos F. (2011b) - Selection of the derivatization reagent: The case of human blood cholesterol, its precursors and phytosterols GC–MS analyses. *Journal of Chromatography B*, 879, 3806 – 3811.

Shaghaghi, M.A., Abumweis, S.S., Jones, P.J.H (2013). Cholesterol-Lowering Efficacy of Plant Sterols/Stanols Provided in Capsule and Tablet Formats: Results of a Systematic

Review and Meta-Analysis. *Journal of the Academy of Nutrition and Dietetics*, doi: 10.1016/j.jand.2013.07.006.

Shanahan, C., Lorimier, R. (2013). Smart Prevention — Health Care Cost Savings Resulting from the Targeted Use of Dietary Supplements: An Economic Case for Promoting Increased Intake of Key Dietary Supplements as a Means to Combat Unsustainable Health Care Cost Growth in the United States. *Frost & Sullivan* [online]. [Cited 2015.05.14]. Available from: http://www.crnusa.org/CRNfoundation /HCCS/chapters /CRNFrostSullivan-fullreport0913.pdf

Sorrentino A.P. (2008) - OTC Product: Centrum Cardio. *Journal of the American Pharmacists Association*, 48, e103 - e104.

Sosinska E., Przybylski R., Hazendonk P., Zhao Y-Y., Curtis J.M. (2013) – Characterization of non-polar dimers formed during thermo-oxidative degradation of β-sitosterol. *Food Chemistry*, 139 (1– 4), 464 – 474.

Srinivasan D., Parkin K. L., Fennema O.R. (2008) – Fennema's Food Chemistry. In McClements D. J., Decker E. A. – Lipids. 4ªedição. London: *Taylor & Francis Group*. ISBN 9780824723453.

Tasan M., Bilgin B., Geçgel U., Demirci A.S. (2006) - Phytosterols as Functional Food Ingredients. *Journal of Tekirdag Agricultural Faculty*, 3 (2), 153 – 159.

Trautwein E., Demonty I. (2007) - Phytosterols: natural compounds with established and emerging health benefits. *Oléagineux, Corps Gras, Lipides*.14, no.5, 259-66 [Cited 2012.06.10]. Available from: http://www.jle.com/

Trautwein, E.A., Duchateau, G.S.M.J.E., Lin, Y., Mel'nikov, S.M., Molhuizen, H.O.F., Ntanios, F.Y. (2003). Proposed mechanisms of cholesterol-lowering action of plant sterols. *European Journal of Lipid Science andTechnology,* 105, 171-185.

Weingartner, O., Böhm, M., Laufs, U. (2009). Controversial Role of Plant Sterol Esters in the Management of Hypercholesterolaemia. *European Heart Journal*, 30(4), 404 – 409.

Winkler-Moser J.K. (2011) - Gas chromatographic analysis of plant sterols. *AOCS Lipid Library*. [Cited 2012.10.28]. Available from: http://lipidlibrary.aocs.org/

Woodgate, D., Chana, C.H.M. and Conquera, J.A. (2006). Cholesterol-Lowering Ability of a Phytostanol Softgel Supplement in Adults with Mild to Moderate Hypercholesterolemia. *Lipids*, 41, 127–132.

Yan F., Yang H., Wu D., Huo M., Li J. (2011) - Recovery of Phytosterols from Waste Residue of Soybean Oil Deodorizer Distillate, Soybean - Applications and Technology, Prof. Tzi-Bun Ng (Ed.) *In Tech*. ISBN: 978-953-307-207-4.

In: Phytosterols
Editor: Deanna Garner

ISBN: 978-1-63483-477-3


*Chapter 2*

# FOOD SOURCES AND CHEMISTRY OF PLANT STEROLS AND STANOLS

***Raquel de Pinho Ferreira Guiné*** *
CI&DETS Research Centre and Department of Food Industry,
Polytechnic Institute of Viseu, ESAV, Quinta da Alagoa,
Estrada de Nelas, Ranhados, Viseu, Portugal

## ABSTRACT

A wide range of plants have phytosterols in their composition, both strerols and stanols, being however the stanols present in considerably smaller quantities when compared to sterols. Among the most abundant plant sterols are sitosterol, campesterol and stigmasterol. As to plant stanols, they are present in the diet in small amounts, being one of the most important the sitostanol.

Some food sources that are rich in phytosterols include for example: fruits, vegetables, nuts, seeds, cereals, legumes and vegetable oils. It is therefore important to know what the food sources of these bioactive compounds are and what amounts are present. This is one of the goals of the present chapter, which aims to revise the most recent scientific literature to look for updates regarding the food sources of phytosterols, as a complement to the information more generally available.

Both stanols and sterols are essential components of plant cell membranes and structurally resemble cholesterol. In fact, cholesterol is also a sterol, but, unlike plant sterols, it is predominantly of animal origin, being synthesized in the human liver. Because the bioactive effects of phytosterols are much associated to their chemical structure, also this aspect is addressed in the present chapter, by presenting the chemistry of the different molecules of the phytosterols.

* E-mail: raquelguine@esav.ipv.pt

# 1. Nature and Roles of Phytosterols

## 1.1. What Are Phytosterols

Phytosterols include sterols and stanols and are natural compounds present in a wide variety of plant sources, which include fruits, vegetables, nuts, seeds, cereals, legumes and vegetable oils, among others. However, the relative amounts are much different, and stanols occur in considerable smaller quantities when compared to sterols (Berger et al., 2004; Piironen et al., 2000). Phytosterols are essential constituents of plant cell membranes and their structure is similar to cholesterol, which is itself also a sterol, but, unlike plant sterols, it is predominantly of animal origin, being synthesized in the human liver. However, some important amounts of cholesterol are also ingested through the diet.

While cholesterol is the major sterol of vertebrates and ergosterol plays a key role in fungi plants usually possess more complex sterol compositions (Dufourc, 2008). Because plant sterols resemble cholesterol, their roles in the human body are also similar. In the human body, strerols and stanols have a potential for lowering blood cholesterol by reducing the absorption of cholesterol by the human intestine. When less dietary and biliary cholesterol is absorbed, the liver increases the removal of harmful cholesterol (LDL: Low-density-lipoproteins) from the blood, thus reducing the serum LDL levels (Richelle et al., 2004). Besides their cholesterol lowering effect, plant sterols also have other positive activities, such as anti-inflammatory, anti-oxidative, anti-atherosclerosis and anti-cancer (Berger et al., 2004).

Although plant stanols and sterols are ingested daily through the diet, the amounts ingested are low. In ancient days, human diets were rich in phytosterols, providing up to 1 g/day. However, the typical Western diet today is relatively low in these plant components and this is why recently there has been an increasing incorporation of such components into food products.

The most abundant plant sterol is sitosterol, followed by campesterol and stigmasterol (Moreau et al., 2002). Also present in the diet in small amounts are plant stanols, such as sitostanol.

Plant stanols and sterols are generally recognized as safe (GRAS). However, they may reduce the absorption of carotenoids (such as beta-carotene) (Gylling et al., 1999; Nguyen et al., 2006), which may be compensated by increasing the intake of food rich in carotenoids, like fruits and vegetables (Noakes et al., 2002).

## 1.2. Role of Phytosterols in Plants

Sterols are a class of lipids that act as regulators in biological processes and sustain the domain structure of cell membranes being considered as membrane reinforcers (Ribeiro et al., 2007; Schaeffer et al., 2001). Stigmasterol (STI) and sitosterol (SIT) are two 24-ethyl sterols, which are major constituents of the sterol profiles in plant species, and are alleged to be involved in the embryonic growth of plants (Dufourc, 2008; Schaller, 2004; Schrick et al., 2002).

Sterols are imperative for the formation of liquid-ordered lipid domains (lipid rafts) which are believed to play an important role in fundamental biological processes, which include signal transduction, cellular sorting, cytoskeleton reorganization and infectious diseases (Simons and Ehehalt, 2002; Simons and Ikonen, 2000). Specialized lipid domains influence the polarized growth of pollen tube and root hair, which are vital for the plants (Dufourc, 2008; Kost et al., 1999).

### 1.3. Role of Phytosterols in Humans

In the human body, phytosterols act similarly to cholesterol, given their structural resemblance. Like cholesterol, phytosterols must be combined with mixed micelles before being caught by enterocytes. After entering the enterocyte, the systemic absorption of phytosterols is inhibited by the activity of efflux transporters, involving of a pair of ATP-binding cassette proteins (Sudhop et al., 2005). These proteins form one half of a transporter that discharges phytosterols as well as unesterified cholesterol from the enterocyte into the intestinal lumen. But because phytosterols are secreted back into the intestine at a much greater rate than cholesterol, this results in a significantly lower intestinal absorption of dietary phytosterols when compared to cholesterol. Within the enterocyte, phytosterols are incorporated into chylomicrons entering the circulation and being taken up by the liver. However, because they are not as readily esterified as it happens with cholesterol, their concentration is considerably lower. Once inside the liver, phytosterols are rapidly secreted into bile by hepatic transporters. Although cholesterol may also be secreted into bile, the rate of phytosterol secretion into bile is much greater than cholesterol secretion (Sudhop et al., 2002). Hence, the lower serum concentrations of phytosterols in relation to cholesterol can be attributed to two effects: the decreased intestinal absorption and the increased excretion of phytosterols into bile.

## 2. CHEMISTRY OF PLANT STANOLS AND STEROLS

Phytosterols are steroid compounds similar in structure to cholesterol and which occur in plant materials varying in carbon side chains and/or presence or absence of a double bond. The general structure of these molecules follows the arrangement shown in Figure 1 that gives origin to the different steroids. More than 200 sterols and related compounds have been reported in literature (Akihisa and Kokke, 1991).

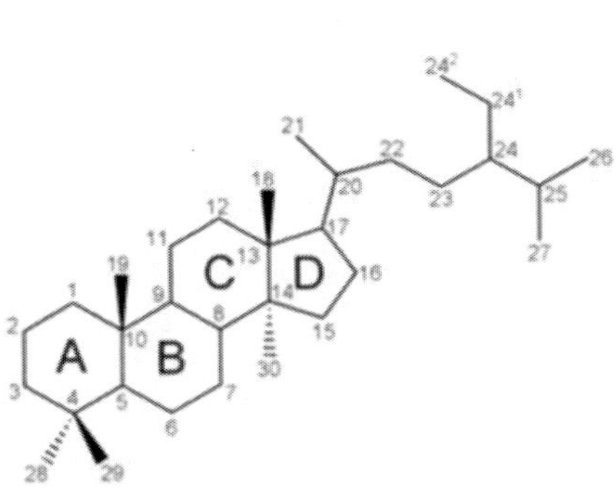


Figure 1. Coding for nomenclature of steroid skeleton.

Stanols correspond to saturated sterols, hence they do not have double bonds in the sterol ring structure.

Isolated phytosterols are insoluble in water, relatively insoluble in oil, and soluble in alcohols.

## 2.1. Plant Sterols

Sterols are a subgroup of steroids with a hydroxyl group bound to the carbon in position 3 of the A-ring, as seen in Figure 2 (Fahy et al., 2005), and giving place to a quite flat molecule. The molecule comprises a polar part (the hydroxyl group on the A ring) and a non-polar part (the rest of the aliphatic chain). Sterols are amphipathic lipids synthesized from acetyl-coenzyme A via the HMG-CoA reductase pathway.

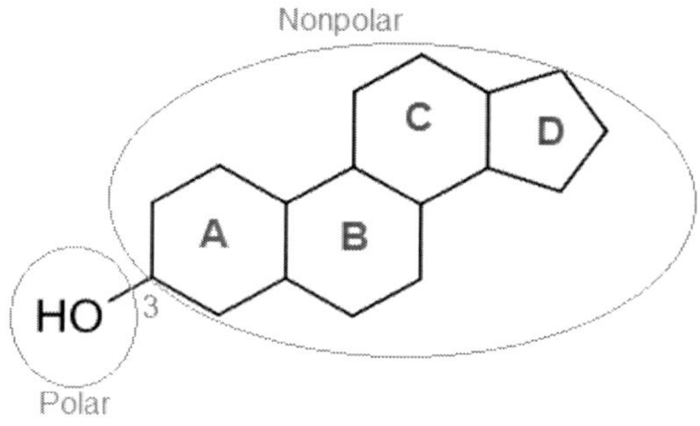


Figure 2. General chemical structure of a sterol.

**Table 1. Name and chemical structure of some sterols**

| | |
|---|---|
| β-Sitosterol 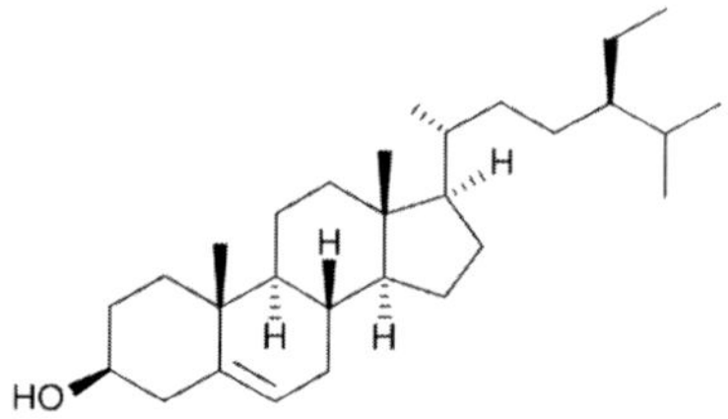 | Campesterol 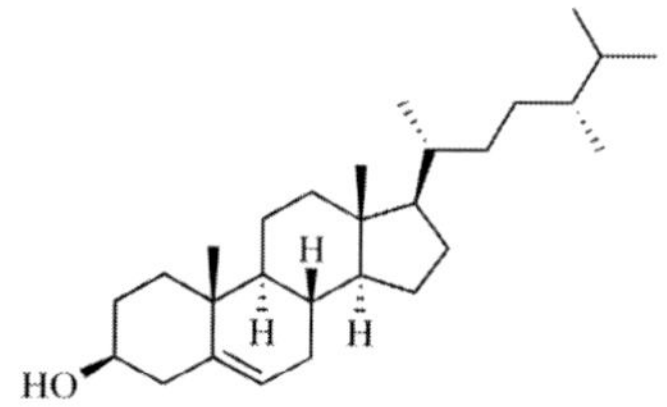 |
| Stigmasterol 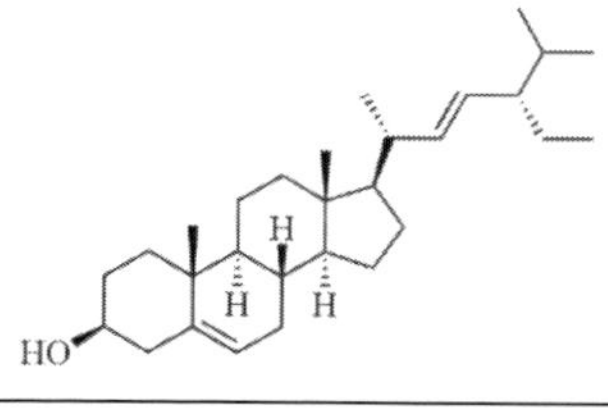 | Brassicasterol 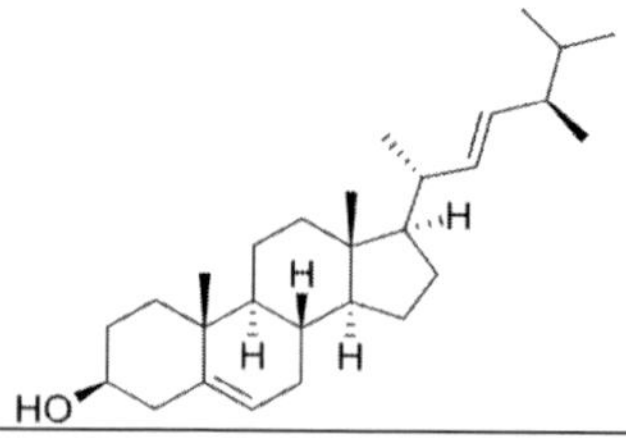 |

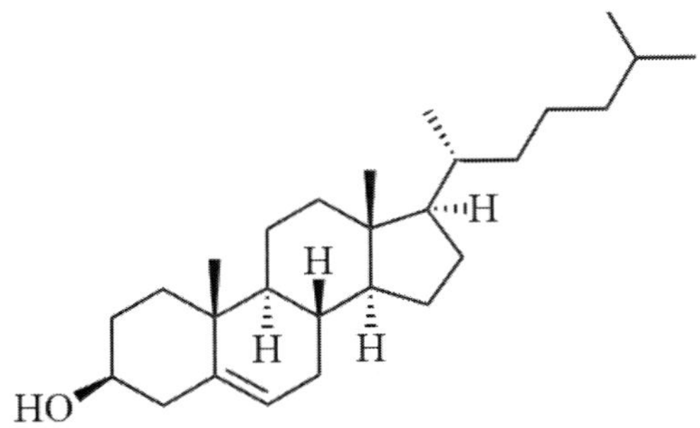

Figure 3. Chemical structure of cholesterol.

As previously mentioned there are a wide range of phytosterols, but some most common and with dietary importance include, among others: sitosterol, campesterol or stigmasterol. Table 1 shows some of the most relevant sterols and Figure 3 shows the chemical structure of cholesterol, a molecule with a common structure, although a different nature.

## 2.2. Plant Stanols

Stanols are saturated sterols, originating from the sterols through hydrogenation. They are also present in the diet, although in small amounts, like for example sitostanol.

**Table 2. Name and chemical structure of some stanols**

| β-sitostanol (Stigmastanol) | Campestanol |
|---|---|
| 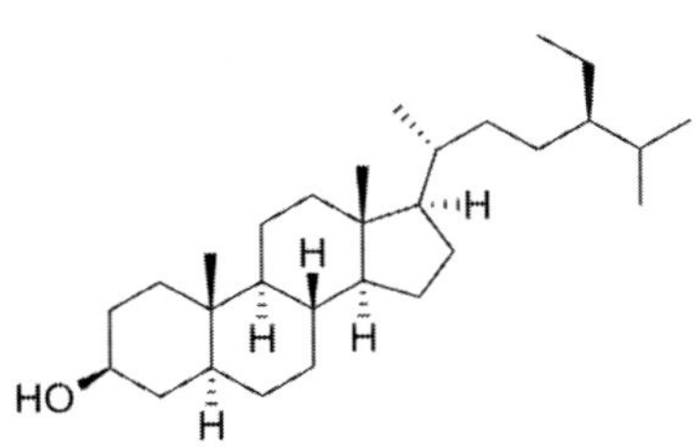 | 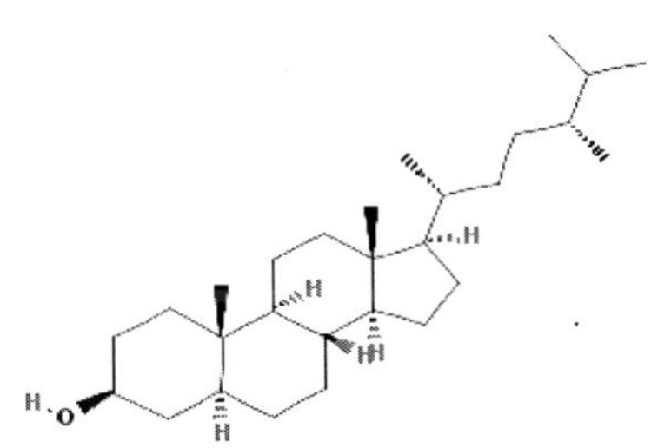 |

# 3. FOOD SOURCES OF PHYTOSTEROLS

It is probable that consumption of plant foods throughout human evolution eventually shaped the dietary requirements of contemporary humans, so that modern man presents a completely different diet and lifestyle. Contrarily to the typical diet in most developed countries today, the diets of our ancestors were rich in phytosterols, providing an estimation of 1000 g/day (Jenkins et al., 2003). Today's intakes of dietary phytosterols are considerable lower than that, being expected to vary from 150-450 mg/day depending on various factors and characteristics of the population (Ostlund, 2002). However, vegetarians, and particularly vegans, generally have the highest intakes of dietary phytosterols. In fact, specially designed vegetarian experimental diets can yield daily intakes of 700 mg/day (Ågren et al., 2001; Nair et al., 1984).

Phytosterols can be present in the free form, which is considered the most biologically active against lowering serum cholesterol. Alternatively, it can also be present as one of four common phytosterol conjugates: (i) fatty acyl esters, (ii) hydroxycinnamate esters, (iii) steryl glycosides, and (iv) fatty acylated steryl glycosides. The bound form is usually hydrolyzed in the small intestines by pancreatic enzymes (Moreau and Hicks, 2004).

Although present in all plant foods, the richest naturally occurring sources of phytosterols are vegetable oils, and their derived products, particularly in unrefined plant oils, including vegetable, nut, and olive oils (Ostlund, 2002). Nuts, seeds, whole grains, and legumes are also good dietary sources of phytosterols (de Jong et al., 2003).

Nuts are rich in phytosterols and can significantly contribute to total phytosterol intake even though eaten in small amounts. On the contrary the daily intakes of cereal products, vegetables, fruit and berries are relatively high, and hence they may also be significant sources of phytosterols even though not much rich in phytosterols (Valsta et al., 2004).

According to Weihrauch and Gardner (1978) the most commonly occurring phytosterols in the human diet are β-sitosterol, campesterol and stigmasterol, representing respectively about 65%, 30% and 3% of diet intakes.

Plant stanol esters, particularly, are found naturally occurring in small quantities in fruits, vegetables, nuts, seeds, cereals, legumes, and vegetable oils. The most common plant stanols in the human diet are sitostanol and campestanol, which combined make up about 5% of dietary phytosterol (Andersson et al., 2004).

Tables 3 and 4 show the overall contents of phytosterols for some plant food products, namely vegetables and fruits (Table 3) and oils and nuts (Table 4).

**Table 3. Phytosterols in some vegetables and fruits (USDA, n.d.)**

| Vegetables | Phytosterols (mg/100 g) | Fruits | Phytosterols (mg/100 g) |
|---|---|---|---|
| Asparagus | 24 | Apple | 12 |
| Beans | 22 | Apricot | 18 |
| Brussels sprouts | 24 | Banana | 16 |
| Cabbage | 11 | Cherrie | 12 |
| Cauliflower | 18 | Fig | 31 |
| Cucumber | 14 | Grapes | 4 |
| Eggplant | 7 | Melon | 10 |
| Green pepper | 9 | Orange | 24 |
| Lettuce (green) | 38 | Peach | 10 |
| Okra | 24 | Pear | 8 |
| Onion | 15 | Pineapple | 6 |
| Potato | 5 | Plum | 7 |
| Pumpkin | 12 | Strawberry | 12 |
| Soybean | 50 | Watermelon | 2 |
| Tomato | 7 | | |

Table 5 details the amounts of some individual sterols, namely stigmasterol, campesterol and β-Sitosterol, in different food sources.

**Table 4. Phytosterols in some oils and nuts (USDA, n.d.)**

| Oils | Phytosterols (mg/100 g) | Nuts | Phytosterols (mg/100 g) |
|---|---|---|---|
| Coconut oil | 86 | Almond | 114 |
| Grapeseed oil | 180 | Chestnut | 22 |
| Olive oil | 221 | Hazelnut | 116 |
| Palm oil | 95 | Macadamia | 116 |
| Peanut oil | 207 | Pistachio | 214 |
| Popyseed oil | 276 | | |
| Rice bran oil | 1190 | | |
| Safflower oil | 444 | | |
| Soy oil | 132 | | |
| Sunflower oil | 100 | | |

**Table 5. Individual sterols in some plant foods (USDA, n.d.)**

| Foods | Stigmasterol (mg/100g) | Campesterol (mg/100g) | β-Sitosterol (mg/100g) |
|---|---|---|---|
| *Oils* | | | |
| Canola oil | 3 | 241 | 413 |
| Corn oil | 56 | 189 | 621 |
| Soybean oil | 59 | 62 | 172 |
| *Nuts* | | | |
| Almond | 4 | 5 | 130 |
| Cashew | 0 | 9 | 113 |
| Hazelnut | 1 | 7 | 102 |
| Macadamia | 0 | 8 | 108 |
| Pecan | 3 | 6 | 117 |
| Pine nut | 0 | 20 | 132 |
| Pistachio | 3 | 10 | 210 |
| Walnut | 0 | 5 | 87 |
| *Vegetables* | | | |
| Beans | 1 | 3 | 18 |
| Pepper | 0 | 1 | 2 |
| Potato | 2 | 3 | 12 |
| *Fruits* | | | |
| Avocado | 2 | 5 | 76 |
| Pomegranate | 0 | 1 | 4 |

# 4. Effects of Phytosterols when Ingested in the Human Diet

## 4.1. Claims

The European Foods Safety Authority (EFSA) experts have concluded that foods such as yoghurt and margarine containing certain levels of plant stanols and sterols can reduce blood cholesterol levels. Scientists on the NDA (Dietetic Products, Nutrition and Allergies) Panel concluded that blood cholesterol can be reduced on average by 7 to 10.5% if a person consumes 1.5 to 2.4 grams of plant sterols and stanols every day. The scientists furthermore observed that that 2–3 weeks are usually enough to establish the desired effect. Longer-term studies extending up to 85 weeks showed that the cholesterol-lowering effect could be sustained (EFSA, 2009).

The EFSA scientists consider that there is evidence supporting the claim that a clinically significant reduction of about 9% in LDL-cholesterol can be achieved by a daily intake of 2 – 2.4 g of phytosterols added to an appropriate food (e.g., plant sterols added to fat-based foods and at foods such as milk and yoghurt). However, the magnitude of the cholesterol-lowering effect may differ among the different food matrices. The Panel concludes that a cause-effect relationship has been established between the consumption of plant sterols and lowering of LDL cholesterol, in a dose-dependent manner (Bresson et al., 2008). The Panel further considers that there is evidence that the risk of coronary heart disease (CHD) can be reduced by cholesterol-lowering therapy including dietary intervention strategies. However, there are no studies demonstrating that plant sterols have an impact on population-based CHD morbidity and mortality rates (Bresson et al., 2008).

Health Canada analysed the evidence provided by 84 randomized controlled trials published between 1994-2007 involving phytosterol supplementation and observed that on average 8.8% of LDL-cholesterol was reduced when a mean intake of 2 grams was made per day. Hence, they concluded that there was sufficient scientific evidence supporting a relationship between phytosterol consumption and blood cholesterol lowering. Based on this evidence Health Canada approved a set of statements for qualifying foods intended for moderately to highly hypercholesterolemic individuals.

The Food and Drug Administration (FDA) has approved the following claims for phytosterols (FDA, 2010, 2005):

- Foods containing at least 0.65 g per serving of plant sterol esters, eaten twice a day with meals for a daily total intake of at least 1.3 g, as part of a diet low in saturated fat and cholesterol, may reduce the risk of heart disease.
- Foods containing at least 1.7 g per serving of plant stanol esters, eaten twice a day with meals for a total daily intake of at least 3.4 g, as part of a diet low in saturated fat and cholesterol, may reduce the risk of heart disease.
- When consumed in the range of 1 to 3 grams in enriched foods, phytosterols resulted in statistically significant (5-15%) reductions in blood LDL cholesterol levels relative to placebo.
- A daily dietary intake of 2 grams a day of phytosterols (expressed as non-esterified phytosterols) is required to make an authorized health claim relating phytosterol consumption to cholesterol lowering and cardiovascular disease (CVD) risk.

## 4.2. Beneficial Physiological Effects

***Lowering of Cholesterol***

No doubt that the ability of phytosterols to reduce cholesterol levels is the most prominent of their benefits for human health, and its proof goes far back to 1953 (Pollak,

1953; Tilvis and Miettinen, 1986). Jones (2007) refers that phytosterols were marketed as a pharmaceutical preparation for the treatment of elevated cholesterol from 1954 to 1982.

Ash et al., (2011) tested whether phytosterol esters containing different sterol moieties (sitosterol, sitostanol, or stigmasterol) but the same fatty acid moiety (stearic acid) produced different effects on cholesterol metabolism. They observed that the consumption of phytosterol esters from the diet in dosages of 2.5% caused only modest impacts on cholesterol metabolism. However, in the case of sitosterol stearate it was found to have a slightly greater therapeutic impact by lowering liver free cholesterol and increasing esterified and total neutral sterol fecal excretion, possibly due to a greater level of intestinal hydrolysis.

Musa-Veloso et al., (2011) compared the LDL-cholesterol lowering efficacy of plant stanols and plant sterols over a continuous dose range, based on 113 publications and 1 unpublished study report. The results showed that consistently stanols were more efficient when compared to sterols. The maximal LDL-cholesterol reductions for plant stanols (16.4%) and plant stanol ester (17.1%) were significantly greater than the maximal LDL-cholesterol reductions for plant sterols (8.3%) and plant sterol ester (8.4%). In view of these results, the authors concluded that intakes of plant stanols over the recommended 2 g/day dose are associated with additional and dose-dependent reductions in LDL-cholesterol, possibly resulting in further reductions in the risk of coronary heart disease (CHD).

Commonly individuals with high HDL-cholesterol (high-density-lipoproteins) have elevated levels of plasma markers of cholesterol absorption (β-sitosterol and campesterol) and correspondingly reduced markers of cholesterol synthesis. The relation between LDL-cholesterol levels and markers of cholesterol absorption is not completely understood, but it has been consistently observed in individuals under a wide range of situations (different plasma cholesterol levels and clinical conditions) (Assmann et al., 2008; Lupattelli et al., 2012; Niesor et al., 2011; Nunes et al., 2011; Wilund et al., 2009).

Shaghaghi et al., (2013) studied the efficacy of plant sterols and stanols in tablet or capsule formats for the lowering of cholesterol as compared to dietary forms. For that they used eight eligible published clinical trials. For intakes of plant sterol/stanol doses ranging from 1.0 to 3.0 g/day administrated mainly with the main meals (2 or 3 times/day), during a period of 4 to 6 weeks, they observed a decrease in LDL-cholesterol concentrations by 12 mg/dL (0.31 mmol/L) as compared with placebo. They also found that the LDL-cholesterol–lowering action of plant sterols/stanols supplements was similar to that of foods enriched with plant sterols/stanols.

### *Lowering of Triglycerides*

Elevated triglyceride levels constitute one of the risk factors for cardiovascular disease (CVD), hence their plasma levels must be under the recommended values (Malloy and Kane, 2001).

Dietary fats are made up mostly of triglyceride with minor lipids such as sterols. Hence, Awad et al., (2000) evaluated the effect of different fatty acids, the main component of triglycerides, and sterols on cell growth and triglyceride accumulation in 3T3-L1 cells, which are used as an in vitro model for studying obesity. The obtained results suggested that diets rich in both monounsaturated fatty acids and phytosterols could play a role in controlling obesity.

Dietary phytosterols reduce LDL cholesterol, and niacin or fenofibrate primarily reduces triglyceride and increases HDL-cholesterol levels. In this way, Yeganeh et al., (2005) hypothesized that a combination of phytosterols with niacin or fenofibrate could synergistically impact lipoprotein profile and atherogenesis in apo E-KO mice. Their results showed that phytosterols alone significantly reduced plasma total cholesterol levels (14.1 vs. 16.9 mmol/L) as well as the extent of atherosclerosis (0.42 vs. 0.15 $mm^2$). The addition of fenofibrate to phytosterols increased plasma total cholesterol levels (14.1 vs. 21.6 mmol/L) and also decreased HDL-cholesterol concentrations (0.8 vs. 0.4 mmol/L). Furthermore, a reduction in the extent of atherosclerosis (0.42 vs. 0.34 $mm^2$) was also observed as compared to controls, suggesting other potential anti-atherogenic effects of fenofibrate.

Plana et al., (2008) observed that in individuals supplemented with 1.6 g/day of plant sterols in a fermented milk beverage for six weeks, the triglycerides decreased by 14%. Plat and Mensik (2009) proposed the mechanism behind the triglyceride-lowering effect of phytosterols as being a reduction in triglyceride-rich VLDL (very-low-density-lipoproteins) produced by the liver. According to Theuwissen et al., (2009) these effects of phytosterols may be particularly noticeable in people with elevated triglycerides.

### *Cancer Prevention and Treatment*

Woyengo et al., (2009) did a review to critically examine results from research regarding the potential effects and mechanisms of action of phytosterols on different forms of cancer. According to their work, considerable emerging evidence supports the inhibitory actions of phytosterols on lung, stomach, as well as ovarian and breast cancer. Furthermore, they evidence the multiple mechanisms of action through which phytosterols seem to act, including inhibition of carcinogen production, cancer-cell growth, angiogenesis, invasion and metastasis, or through the promotion of apoptosis of cancerous cells.

It has been reported that β-sitosterol, the prevalent phytosterol in the diet, has anti-proliferative effects in colon cancer in rats and in cell lines (Baskar et al., 2012; Cilla et al., 2015).

According to the review made by Grattan (2013), the cholesterol modulating actions of plant sterols may overlap with their anti-cancer actions. Breast cancer is the most common malignancy affecting women and there remains a need for effective adjuvant therapies for this disease, for which plant sterols may play a distinctive role. Llaverias et al., (2013) in their study provide preclinical proof that dietary phytosterols could prevent the breast tumor growth associated with fat-rich diet consumption.

Essays in vitro demonstrated a role of phytosterols on prostate cancer. They were observed to indirectly (in vivo as a dietary supplement) and directly (in tissue culture media) inhibit the growth and metastasis of PC-3 cells. Furthermore, beta-sitosterol was more effective than campesterol in offering this protection (Awad et al., 2001). Ifere et al., (2009) concluded that sterol intake may contribute to the disparity in incidence of prostate cancer. Also they found that the knowledge of the mechanism for modulation of growth and apoptosis signaling may reveal potential targets for cancer prevention and/or chemotherapeutic intervention. Mendilaharsu et al., (1998) conducted a case control study in Uruguay to evaluate the role of phytosterols in diminishing the risk of lung cancer. They observed a 50% reduction in the risk of lung cancer among the individuals with the highest intake of phytosterols.

In a case-control work by De Stefani et al., (2000) which assessed the effects of phytosterols on stomach cancer risk, the results showed an odds ratio of 0.33 among those with the highest phytosterols intake. At present some epidemiological data suggests a correlation between plant sterol intake and a reduction in cancer risk, and it has been estimated that phytochemical intake may be related to a reduction in cancer risk upwards of 20% (Grattan, 2013).

***Neurodegenerative Diseases***

Oxidative stress and apoptosis are frequently associated with neuronal cell damage in various neurodegenerative disorders, such as Parkinson's disease. Brassinosteroids are phytosterols recognized to promote stress tolerance of vegetables via modulation of the antioxidative enzyme cascade. Hence Carange et al., (2011) analyzed the ability of a natural brassinosteroid (24-epibrassinolide) to protect neuronal PC12 cells from apoptosis. Their results demonstrated the potent antioxidant and neuroprotective role of 24-epibrassinolide in a mammalian neuronal cell line.

Alzheimer's disease is the most prevalent progressive neurodegenerative disorder in elderly. Phytosterols have been shown to beneficially modulate molecular processes involved in this disease (Vanmierlo et al., 2015).

## 4.3. Safety

The safety of stanols and sterols has been reviewed by several regulatory agencies. Both plant sterols and stanols are generally recognized as safe (GRAS) by the FDA (USA) (FDA, 2003). Also the Scientific Committee on Foods of the EU concluded that plant sterols and stanols are safe for human use (SCF-EU, 2003). Hence, both organisms state that the addition of plant sterols and stanols to foods, as a functioning component, does not impart in any kind of harm for the human health. However, the Committee made some recommendations regarding the maximum daily intake of sterols and stanols that should not exceed 3 g/day. This limit was established because on one hand higher intakes might cause undesirable effects and on the other hand there is no evidence of increased health benefits.

## 4.4. Adverse Effects

Hendricks et al., (2003) reported that people who consumed a plant sterol-enriched spread providing 1.6 g/day did not report any more adverse effects than those consuming a control spread for up to one year, and people consuming a plant stanol-enriched spread providing 1.8-2.6 g/day for the same period did not report any adverse effects at all.

Ayesh et al., (1999) conducted a study to determine the effect of a margarine enriched with phytosterol esters on faecal short-chain fatty acids and faecal bacterial enzyme activities, viable faecal microflora count, female sex hormones and serum cholesterol concentrations in healthy individuals. They concluded that an ingestion of up to 8.6 g/day of phytosterols in margarine for three to four weeks was well-tolerated and did not adversely affect intestinal bacteria or female hormone levels.

In general no adverse effects have been associated with regular consumption of plant sterols or stanols. Still, and although phytosterols are usually well-tolerated, some undesirable effects have occasionally been reported, such as nausea, indigestion, diarrhoea, and constipation (Berges et al., 1995; Klippel et al., 1997).

***Sitosterolemia***

Sitosterolemia, also known as phytosterolemia, is a rare, autosomal recessive lipid disorder, which is characterized by elevated plasma levels of plant sterols due to increased intestinal absorption and reduced biliary secretion of neutral sterols. Patients with sitosterolemia are frequently susceptible to developing a number of pathologies: hypercholesterolemia, xanthomas, premature coronary heart disease, hemolysis, arthralgias and arthritis. Sitosterolemia results from inheriting a mutation in both copies of the ABCG5 or ABCG8 genes (Berge, 2003). Individuals with this pathology have preoccupying elevated concentrations of phytosterols in the blood due to increased intestinal absorption and decreased biliary excretion. Hence, they are at high risk for premature atherosclerosis, and therefore should avoid foods or supplements with added plant sterols (Berger et al., 2004).

Stalenhoef et al., (2001) examined the effect of plant sterol-enriched margarine on plasma lipids and sterols in subjects heterozygous for phytosterolaemia. Their results showed that an intake of around 3 g/day of plant sterols by subjects heterozygous for phytosterolaemia increased campesterol or sitosterol levels in blood to similar levels as found in normal subjects. In addition, plasma cholesterol levels were reduced to the same extent as in normal or hypercholesterolaemic individuals. Kwiterovich et al., (2003) in a similar work concluded that the consumption of 2.2 g/day of plant sterols for 6-12 weeks by heterozygous carriers did not result in abnormally elevated serum phytosterols.

Othman et al., (2013) did a review with the objective of evaluating sterol metabolism in Sitosterolemia and the impact of the currently available treatments on sterol trafficking in this disease. Accordng to their findings, blocking sterols absorption with ezetimibe has revolutionized the treatment of these patients because it reduces circulating levels of non-cholesterol sterols. However, none of the available treatments including ezetimibe have normalized plasma plant sterols concentrations.

***Interaction with Carotenoids***

Dietary carotenoids are fat-soluble phytochemicals that circulate in lipoproteins. Substances that can reduce the circulating carotenoids may be a serious concern because low levels of circulating carotenoids are reported to be associated with increased risk of several chronic diseases like CHD, certain types of cancer or macular degeneration (Katan et al., 2003; Kritchevsky, 1999).

Katan et al., (2003) evaluated the efficacy and safety of plant stanols and sterols in the management of blood cholesterol levels. They analysed many published studies and found a number of them where reductions in plasma carotenoids were observed, varying between 10 and 20%, after short-term and long-term consumption of foods enriched with plant sterols or stanols. Plat and Mensink (2001) studied the effects of diets enriched with plant stanol ester mixtures on plasma fat-soluble antioxidant concentrations. They observed persisting decreases in concentration of α-carotene, β-carotene, and lycopene, suggesting that phytosterols can inhibit the absorption of these carotenoids. Ntanios and Duchateau (2002) made a review about carotenoids and the use of plant sterol-enriched spreads. They found that it is generally recognized that plant sterol/stanol-enriched, cholesterol-lowering spreads, if consumed regularly, may induce a 10-20% decrease in plasma carotenoids. However, according to them, a 10-20% decrease in plasma carotenoids falls well within the seasonal variation observed in individuals, and hence their understanding is that the slight decrease of carotenoid blood levels due to the intake of plant phytosterols does not point to any health risk. On the other hand, several studies have found that increasing intakes of carotenoid-rich fruit and vegetables can prevent phytosterol-induced decreases in plasma carotenoids (Ntanios and Duchateau, 2002). Noakes et al., (2002) defend that an increase in dietary carotenoids when consuming plant sterols or stanols is effective in maintaining plasma carotenoid concentrations. In particular their advice is to consume five daily servings of fruit and vegetables, including one serving of carotenoid-rich vegetables, as being enough to maintain plasma carotenoid levels in people consuming 2.5 g/day of plant sterol or stanol esters.

***Interaction with Fat-Soluble Vitamins***

Because plant sterols and stanols decrease cholesterol absorption and serum LDL cholesterol concentrations, their effects on fat-soluble vitamin concentrations have also been studied. In general, the studies have revealed that the consumption of foods enriched with plant sterols and stanols at doses of 1.5 g/day, or more, have not been correlated with adverse effects on blood fat-soluble vitamin concentrations in well-nourished populations.

According to Katan et al., (2003) a meta-analysis of 10 to 15 trials showed that plasma levels of vitamin A (retinol) are not affected by stanols or sterols. Also Hendriks et al., (2003) observed that vitamin A concentrations were not affected by plant stanol or sterol ester consumption for up to one year. Although the majority of studies found no changes in plasma vitamin D (25-hydroxyvitamin D3) concentrations (Katan et al., 2003), one placebo-controlled study in individuals consuming 1.6 g/day of sterol esters for one year observed a small (7%) but statistically significant decrease in plasma vitamin D concentrations (Hendriks et al., 2003). Consumption of plant sterol or stanol-enriched foods has been found to decrease plasma vitamin E (α-tocopherol) concentrations in a number published works (Katan et al., 2003; Korpela et al., 2006). Nevertheless, generally when plasma α-tocopherol concentrations are standardized to LDL cholesterol concentrations, those unbalances disappear. It is not evident that plant sterol or stanol consumption adversely affects vitamin K concentrations. Although a consumption of 1.6 g/day of sterol esters for six months was associated with a decrease of 14 % in plasma vitamin K1 concentrations, this was not significant because carboxylated osteocalcin, a functional indicator of vitamin K status, was not affected (Hendriks et al., 2003). In other studies of shorter duration, consumption of plant sterol and stanol esters did not significantly change plasma concentrations of vitamin K1(Korpela et al., 2006; Raeini-Sarjaz et al., 2002) or vitamin K-dependent clotting factors (Plat and Mensink, 2000).

## CONCLUSION

Phytosterols include sterols and stanols, which are similar in structure to cholesterol, being also absorbed through the same mechanisms. All these molecules share a common structure, but while sterols are unsaturated, the stanols are saturated and hence do not have double bonds in the structure of the aromatic rings.

A diet rich in phytosterols can contribute for the absorption of a lower amount of cholesterol, which can lead to lower blood LDL (or, "bad") cholesterol levels and a reduced risk of heart disease. Furthermore, it has been demonstrated the role of phytosterols in helping prevent or fighting several types of cancer, as well as different neurodegenerative disorders, such as Parkinson's or Alzheimer's.

## Acknowledgments

The author thanks the valuable contributions of the reviewers of the present chapter:

Prof. Maria João Barroca (PhD), Molecular Physical-Chemistry Group, Coimbra University Research Centre.

Prof. Luísa Paula Valente (PhD), Department of Environment, Polytechnic Institute of Viseu, Portugal.

## References

Ågren, J.J., Tvrzicka, E., Nenonen, M.T., Helve, T., Hänninen, O., 2001. Divergent changes in serum sterols during a strict uncooked vegan diet in patients with rheumatoid arthritis. *British Journal of Nutrition* 85, 137–139. doi:10.1079/BJN2000234

Akihisa, T., Kokke, W., 1991. Naturally occurring sterols and related compounds from plants, in: Physiology and Biochemistry of Sterols. Champaign, IL: American Oil Chemists' Society, pp. 172–228.

Andersson, S.W., Skinner, J., Ellegård, L., Welch, A.A., Bingham, S., Mulligan, A., Andersson, H., Khaw, K.-T., 2004. Intake of dietary plant sterols is inversely related to serum cholesterol concentration in men and women in the EPIC Norfolk population: a cross-sectional study. *Eur. J. Clin. Nutr.* 58, 1378–1385. doi:10.1038/sj.ejcn.1601980

Ash, M.M., Hang, J., Dussault, P.H., Carr, T.P., 2011. Phytosterol stearate esters elicit similar responses on plasma lipids and cholesterol absorption but different responses on fecal neutral sterol excretion and hepatic free cholesterol in male Syrian hamsters. *Nutrition Research* 31, 537–543. doi:10.1016/j.nutres.2011.06.007

Assmann, G., Kannenberg, F., Ramey, D.R., Musliner, T.A., Gutkin, S.W., Veltri, E.P., 2008. Effects of ezetimibe, simvastatin, atorvastatin, and ezetimibe-statin therapies on non-cholesterol sterols in patients with primary hypercholesterolemia. *Curr. Med. Res. Opin.* 24, 249–259. doi:10.1185/030079908X253663

Awad, A.B., Begdache, L.A., Fink, C.S., 2000. Effect of sterols and fatty acids on growth and triglyceride accumulation in 3T3-L1 cells. The Journal of *Nutritional Biochemistry* 11, 153–158. doi:10.1016/S0955-2863(99)00087-X

Awad, A.B., Fink, C.S., Williams, H., Kim, U., 2001. In vitro and in vivo (SCID mice) effects of phytosterols on the growth and dissemination of human prostate cancer PC-3 cells. *Eur. J. Cancer Prev.* 10, 507–513.

Ayesh, R., Weststrate, J.A., Drewitt, P.N., Hepburn, P.A., 1999. Safety evaluation of phytosterol esters. Part 5. Faecal short-chain fatty acid and microflora content, faecal bacterial enzyme activity and serum female sex hormones in healthy normolipidaemic volunteers consuming a controlled diet either with or without a phytosterol ester-enriched margarine. *Food Chem. Toxicol.* 37, 1127–1138.

Baskar, A.A., Al Numair, K.S., Gabriel Paulraj, M., Alsaif, M.A., Muamar, M.A., Ignacimuthu, S., 2012. β-sitosterol prevents lipid peroxidation and improves antioxidant status and histoarchitecture in rats with 1,2-dimethylhydrazine-induced colon cancer. *J. Med. Food* 15, 335–343. doi:10.1089/jmf.2011.1780

Berge, K.E., 2003. Sitosterolemia: a gateway to new knowledge about cholesterol metabolism. Ann. Med. 35, 502–511.

Berger, A., Jones, P.J., Abumweis, S.S., 2004. Plant sterols: factors affecting their efficacy and safety as functional food ingredients. *Lipids Health Dis* 3, 5. doi:10.1186/1476-511X-3-5

Berges, R.R., Windeler, J., Trampisch, H.J., Senge, T., 1995. Randomised, placebo-controlled, double-blind clinical trial of beta-sitosterol in patients with benign prostatic hyperplasia. Beta-sitosterol Study Group. *Lancet* 345, 1529–1532.

Bresson, J.-L., Flynn, A., Heinonen, M., Hulshof, K., Korhonen, H., Lagiou, P., Løvik, M., Marchelli, R., Martin, A., Moseley, B., Palou, A., Przyrembel, H., Salminnen, S., Strain, S., Strobel, S., Tetens, I., Berg, H., Loveren, H., Verhagen, H., 2008. Plant Sterols and Blood Cholesterol. Scientific substantiation of a health claim related to plant sterols and lower/reduced blood cholesterol and redu ced risk of (coronary) heart disease pursuant to Article 14 of Regulation (EC) No 1924/2006. Scientific Opinion of the Panel on Diet etic Products, Nutrition and Allergies. *The EFSA Journal* 781, 1–12.

Carange, J., Longpré, F., Daoust, B., Martinoli, M.-G., 2011. 24-Epibrassinolide, a Phytosterol from the Brassinosteroid Family, Protects Dopaminergic Cells against MPP+-Induced Oxidative Stress and Apoptosis. *J. Toxicol* 2011. doi:10.1155/2011/392859

Cilla, A., Attanzio, A., Barberá, R., Tesoriere, L., Livrea, M.A., 2015. Anti-proliferative effect of main dietary phytosterols and β-cryptoxanthin alone or combined in human colon cancer Caco-2 cells through cytosolic Ca+2 – and oxidative stress-induced apoptosis. *Journal of Functional Foods* 12, 282–293. doi:10.1016/j.jff.2014.12.001

De Jong, A., Plat, J., Mensink, R.P., 2003. Metabolic effects of plant sterols and stanols (Review). *J. Nutr. Biochem.* 14, 362–369.

De Stefani, E., Boffetta, P., Ronco, A.L., Brennan, P., Deneo-Pellegrini, H., Carzoglio, J.C., Mendilaharsu, M., 2000. Plant sterols and risk of stomach cancer: a case-control study in Uruguay. *Nutr Cancer* 37, 140–144. doi:10.1207/S15327914NC372_4

Dufourc, E.J., 2008. The role of phytosterols in plant adaptation to temperature, in: *Plant Signaling & Behavior*. pp. 133–134.

EFSA, 2009. EFSA News Story: Blood cholesterol reduction health claims on phytosterols can now be judged against EFSA new scientific advice.

Fahy, E., Subramaniam, S., Brown, H.A., Glass, C.K., Merrill, A.H., Murphy, R.C., Raetz, C.R.H., Russell, D.W., Seyama, Y., Shaw, W., Shimizu, T., Spener, F., Meer, G. van, VanNieuwenhze, M.S., White, S.H., Witztum, J.L., Dennis, E.A., 2005. A comprehensive classification system for lipids. *J. Lipid Res.* 46, 839–862. doi:10.1194/jlr.E400004-JLR200

FDA, 2010. Food Labeling; Health Claim; Phytosterols and Risk of Coronary Heart Disease; Proposed Rule (Department of Health and Human Services No. Part II - 21 CFR Part 101), Department of Health and Human Services. *Food and Drug Administration.*

FDA, 2005. Health claims: plant sterol/stanol esters and risk of coronary heart disease (CHD). (Electronic Code of Federal Regulations No. §101.83), Title 21: Food and Drugs. Part

101-Food Labeling. Subpart E-Specific Requirements for Health Claims. *Food and Drug Administration*.

FDA, 2003. Food and Drug Administration. GRAS Notice No. GRN 000112.

Grattan, B.J., 2013. Plant Sterols as Anticancer Nutrients: Evidence for Their Role in Breast Cancer. *Nutrients* 5, 359–387. doi:10.3390/nu5020359

Gylling, H., Puska, P., Vartiainen, E., Miettinen, T.A., 1999. Retinol, vitamin D, carotenes and alpha-tocopherol in serum of a moderately hypercholesterolemic population consuming sitostanol ester margarine. *Atherosclerosis* 145, 279–285.

Hendriks, H.F.J., Brink, E.J., Meijer, G.W., Princen, H.M.G., Ntanios, F.Y., 2003. Safety of long-term consumption of plant sterol esters-enriched spread. *Eur. J. Clin. Nutr.* 57, 681–692. doi:10.1038/sj.ejcn.1601598

Ifere, G.O., Barr, E., Equan, A., Gordon, K., Singh, U.P., Chaudhary, J., Igietseme, J.U., Ananaba, G.A., 2009. Differential effects of cholesterol and phytosterols on cell proliferation, apoptosis and expression of a prostate specific gene in prostate cancer cell lines. *Cancer Detection and Prevention* 32, 319–328. doi:10.1016/j.cdp.2008.12.002

Jenkins, D.J.A., Kendall, C.W.C., Marchie, A., Jenkins, A.L., Connelly, P.W., Jones, P.J.H., Vuksan, V., 2003. The Garden of Eden--plant based diets, the genetic drive to conserve cholesterol and its implications for heart disease in the 21st century. *Comp. Biochem. Physiol., Part A Mol. Integr. Physiol.* 136, 141–151.

Jones, P.J.H., 2007. Ingestion of phytosterols is not potentially hazardous. *J. Nutr.* 137, 2485; author reply 2486.

Katan, M.B., Grundy, S.M., Jones, P., Law, M., Miettinen, T., Paoletti, R., Stresa Workshop Participants, 2003. Efficacy and safety of plant stanols and sterols in the management of blood cholesterol levels. *Mayo Clin. Proc.* 78, 965–978. doi:10.4065/78.8.965

Klippel, K.F., Hiltl, D.M., Schipp, B., 1997. A multicentric, placebo-controlled, double-blind clinical trial of beta-sitosterol (phytosterol) for the treatment of benign prostatic hyperplasia. German BPH-Phyto Study group. *Br. J. Urol.* 80, 427–432.

Korpela, R., Tuomilehto, J., Högström, P., Seppo, L., Piironen, V., Salo-Väänänen, P., Toivo, J., Lamberg-Allardt, C., Kärkkäinen, M., Outila, T., Sundvall, J., Vilkkilä, S., Tikkanen, M.J., 2006. Safety aspects and cholesterol-lowering efficacy of low fat dairy products containing plant sterols. *Eur. J. Clin. Nutr.* 60, 633–642. doi:10.1038/sj.ejcn.1602362

Kost, B., Lemichez, E., Spielhofer, P., Hong, Y., Tolias, K., Carpenter, C., Chua, N.H., 1999. Rac homologues and compartmentalized phosphatidylinositol 4, 5-bisphosphate act in a common pathway to regulate polar pollen tube growth. *J. Cell Biol.* 145, 317–330.

Kritchevsky, S.B., 1999. beta-Carotene, carotenoids and the prevention of coronary heart disease. *J. Nutr.* 129, 5–8.

Kwiterovich, P.O., Chen, S.C., Virgil, D.G., Schweitzer, A., Arnold, D.R., Kratz, L.E., 2003. Response of obligate heterozygotes for phytosterolemia to a low-fat diet and to a plant sterol ester dietary challenge. *J. Lipid Res.* 44, 1143–1155. doi:10.1194/jlr.M200455-JLR200

Llaverias, G., Escolà-Gil, J.C., Lerma, E., Julve, J., Pons, C., Cabré, A., Cofán, M., Ros, E., Sánchez-Quesada, J.L., Blanco-Vaca, F., 2013. Phytosterols inhibit the tumor growth and lipoprotein oxidizability induced by a high-fat diet in mice with inherited breast cancer. *The Journal of Nutritional Biochemistry* 24, 39–48. doi:10.1016/j.jnutbio.2012.01.007

Lupattelli, G., Pirro, M., Siepi, D., Mannarino, M.R., Roscini, A.R., Vaudo, G., Pasqualini, L., Schillaci, G., Mannarino, E., 2012. Non-cholesterol sterols in different forms of primary hyperlipemias. *Nutr. Metab. Cardiovasc Dis* 22, 231–236. doi:10.1016/j.numecd.2010.05.010

Malloy, M.J., Kane, J.P., 2001. A risk factor for atherosclerosis: triglyceride-rich lipoproteins. *Adv. Intern Me.d* 47, 111–136.

Mendilaharsu, M., De Stefani, E., Deneo-Pellegrini, H., Carzoglio, J., Ronco, A., 1998. Phytosterols and risk of lung cancer: A case-control study in Uruguay. *Lung Cancer* 21, 37–45. doi:10.1016/S0169-5002(98)00044-0

Moreau, R.A., Hicks, K.B., 2004. The in vitro hydrolysis of phytosterol conjugates in food matrices by mammalian digestive enzymes. *Lipids* 39, 769–776.

Moreau, R.A., Whitaker, B.D., Hicks, K.B., 2002. Phytosterols, phytostanols, and their conjugates in foods: structural diversity, quantitative analysis, and health-promoting uses. *Prog. Lipid Res.* 41, 457–500.

Musa-Veloso, K., Poon, T.H., Elliot, J.A., Chung, C., 2011. A comparison of the LDL-cholesterol lowering efficacy of plant stanols and plant sterols over a continuous dose range: Results of a meta-analysis of randomized, placebo-controlled trials. *Prostaglandins, Leukotrienes and Essential Fatty Acids* 85, 9–28. doi:10.1016/j.plefa.2011.02.001

Nair, P.P., Turjman, N., Kessie, G., Calkins, B., Goodman, G.T., Davidovitz, H., Nimmagadda, G., 1984. Diet, nutrition intake, and metabolism in populations at high and low risk for colon cancer. Dietary cholesterol, beta-sitosterol, and stigmasterol. *Am. J. Clin. Nutr.* 40, 927–930.

Nguyen, K.A., Cao, Y., Chen, J.R., Townsend, C.M., Ko, T.C., 2006. Dietary Fiber Enhances a Tumor Suppressor Signaling Pathway in the Gut. *Ann. Surg.* 243, 619–627. doi:10.1097/01.sla.0000216783.85214.c1

Niesor, E.J., Chaput, E., Staempfli, A., Blum, D., Derks, M., Kallend, D., 2011. Effect of dalcetrapib, a CETP modulator, on non-cholesterol sterol markers of cholesterol homeostasis in healthy subjects. *Atherosclerosis* 219, 761–767. doi:10.1016/j.atherosclerosis.2011.09.017

Noakes, M., Clifton, P., Ntanios, F., Shrapnel, W., Record, I., McInerney, J., 2002. An increase in dietary carotenoids when consuming plant sterols or stanols is effective in maintaining plasma carotenoid concentrations. *Am. J. Clin. Nutr.* 75, 79–86.

Ntanios, F.Y., Duchateau, G.S.M.J.E., 2002. A healthy diet rich in carotenoids is effective in maintaining normal blood carotenoid levels during the daily use of plant sterol-enriched spreads. *Int. J. Vitam Nutr. Res.* 72, 32–39.

Nunes, V.S., Leança, C.C., Panzoldo, N.B., Parra, E., Cazita, P.M., Nakandakare, E.R., de Faria, E.C., Quintão, E.C.R., 2011. HDL-C concentration is related to markers of absorption and of cholesterol synthesis: Study in subjects with low vs. high HDL-C. *Clinica Chimica Acta* 412, 176–180. doi:10.1016/j.cca.2010.09.039

Ostlund, R.E., 2002. Phytosterols in human nutrition. *Annu. Rev. Nutr.* 22, 533–549. doi:10.1146/annurev.nutr.22.020702.075220

Othman, R.A., Myrie, S.B., Jones, P.J.H., 2013. Non-cholesterol sterols and cholesterol metabolism in sitosterolemia. *Atherosclerosis* 231, 291–299. doi:10.1016/j.atherosclerosis.2013.09.038

Piironen, V., Lindsay, D.G., Miettinen, T.A., Toivo, J., Lampi, A.-M., 2000. Plant sterols: biosynthesis, biological function and their importance to human nutrition. *J. Sci. Food*

*Agric.* 80, 939–966. doi:10.1002/(SICI)1097-0010(20000515)80:7<939::AID-JSFA644>3.0.CO;2-C

Plana, N., Nicolle, C., Ferre, R., Camps, J., Cos, R., Villoria, J., Masana, L., DANACOL group, 2008. Plant sterol-enriched fermented milk enhances the attainment of LDL-cholesterol goal in hypercholesterolemic subjects. *Eur. J. Nutr.* 47, 32–39. doi:10.1007/s00394-007-0693-4

Plat, J., Mensink, R.P., 2009. Plant Stanol Esters Lower Serum Triacylglycerol Concentrations via a Reduced Hepatic VLDL-1 Production. *Lipids* 44, 1149–1153. doi:10.1007/s11745-009-3361-z

Plat, J., Mensink, R.P., 2001. Effects of diets enriched with two different plant stanol ester mixtures on plasma ubiquinol-10 and fat-soluble antioxidant concentrations. *Metab. Clin. Exp.* 50, 520–529. doi:10.1053/meta.2001.22509

Plat, J., Mensink, R.P., 2000. Vegetable oil based versus wood based stanol ester mixtures: effects on serum lipids and hemostatic factors in non-hypercholesterolemic subjects. *Atherosclerosis 1*48, 101–112.

Pollak, O.J., 1953. Reduction of blood cholesterol in man. *Circulation* 7, 702–706.

Raeini-Sarjaz, M., Ntanios, F.Y., Vanstone, C.A., Jones, P.J.H., 2002. No changes in serum fat-soluble vitamin and carotenoid concentrations with the intake of plant sterol/stanol esters in the context of a controlled diet. *Metab. Clin. Exp.* 51, 652–656.

Ribeiro, N., Streiff, S., Heissler, D., Elhabiri, M., Albrecht-Gary, A.M., Atsumi, M., Gotoh, M., Désaubry, L., Nakatani, Y., Ourisson, G., 2007. Reinforcing effect of bi- and tri-cyclopolyprenols on "primitive" membranes made of polyprenyl phosphates. *Tetrahedron* 63, 3395–3407. doi:10.1016/j.tet.2007.01.076

Richelle, M., Enslen, M., Hager, C., Groux, M., Tavazzi, I., Godin, J.-P., Berger, A., Métairon, S., Quaile, S., Piguet-Welsch, C., Sagalowicz, L., Green, H., Fay, L.B., 2004. Both free and esterified plant sterols reduce cholesterol absorption and the bioavailability of beta-carotene and alpha-tocopherol in normocholesterolemic humans. *Am. J. Clin. Nutr.* 80, 171–177.

SCF-EU, 2003. Scientific Committee on Food. Opinion on Applications for Approval of a Variety of Plant Sterol-Enriched Foods.

Schaeffer, A., Bronner, R., Benveniste, P., Schaller, H., 2001. The ratio of campesterol to sitosterol that modulates growth in Arabidopsis is controlled by STEROL METHYLTRANSFERASE 2;1. *Plant J.* 25, 605–615.

Schaller, H., 2004. New aspects of sterol biosynthesis in growth and development of higher plants. *Plant Physiol. Biochem.* 42, 465–476. doi:10.1016/j.plaphy.2004.05.012

Schrick, K., Mayer, U., Martin, G., Bellini, C., Kuhnt, C., Schmidt, J., Jürgens, G., 2002. Interactions between sterol biosynthesis genes in embryonic development of Arabidopsis. *Plant J. 31*, 61–73.

Shaghaghi, M.A., Abumweis, S.S., Jones, P.J.H., 2013. Cholesterol-Lowering Efficacy of Plant Sterols/Stanols Provided in Capsule and Tablet Formats: Results of a Systematic Review and Meta-Analysis. *Journal of the Academy of Nutrition and Dietetics* 113, 1494–1503. doi:10.1016/j.jand.2013.07.006

Simons, K., Ehehalt, R., 2002. Cholesterol, lipid rafts, and disease. *J. Clin. Invest.* 110, 597–603. doi:10.1172/JCI16390

Simons, K., Ikonen, E., 2000. How cells handle cholesterol. *Science* 290, 1721–1726.

Stalenhoef, A.F., Hectors, M., Demacker, P.N., 2001. Effect of plant sterol-enriched margarine on plasma lipids and sterols in subjects heterozygous for phytosterolaemia. *J. Intern. Med.* 249, 163–166.

Sudhop, T., Lütjohann, D., von Bergmann, K., 2005. Sterol transporters: targets of natural sterols and new lipid lowering drugs. *Pharmacol. Ther.* 105, 333–341. doi:10.1016/j.pharmthera.2004.10.011

Sudhop, T., Sahin, Y., Lindenthal, B., Hahn, C., Lüers, C., Berthold, H.K., von Bergmann, K., 2002. Comparison of the hepatic clearances of campesterol, sitosterol, and cholesterol in healthy subjects suggests that efflux transporters controlling intestinal sterol absorption also regulate biliary secretion. *Gut* 51, 860–863.

Theuwissen, E., Plat, J., van der Kallen, C.J., van Greevenbroek, M.M., Mensink, R.P., 2009. Plant stanol supplementation decreases serum triacylglycerols in subjects with overt hypertriglyceridemia. *Lipids* 44, 1131–1140. doi:10.1007/s11745-009-3367-6

Tilvis, R.S., Miettinen, T.A., 1986. Serum plant sterols and their relation to cholesterol absorption. *Am. J. Clin. Nutr.* 43, 92–97.

USDA, n.d. USDA National Nutrient Database for Standard Reference.

Valsta, L.M., Lemström, A., Ovaskainen, M.-L., Lampi, A.-M., Toivo, J., Korhonen, T., Piironen, V., 2004. Estimation of plant sterol and cholesterol intake in Finland: quality of new values and their effect on intake. *Br. J. Nutr.* 92, 671–678.

Vanmierlo, T., Bogie, J.F.J., Mailleux, J., Vanmol, J., Lütjohann, D., Mulder, M., Hendriks, J.J.A., 2015. Plant sterols: Friend or foe in CNS disorders? *Progress in Lipid Research* 58, 26–39. doi:10.1016/j.plipres.2015.01.003

Weihrauch, J.L., Gardner, J.M., 1978. Sterol content of foods of plant origin. *J. Am. Diet Assoc.* 73, 39–47.

Wilund, K.R., Feeney, L.A., Tomayko, E.J., Weiss, E.P., Hagberg, J.M., 2009. Effects of endurance exercise training on markers of cholesterol absorption and synthesis. *Physiol. Res.* 58, 545–552.

Woyengo, T.A., Ramprasath, V.R., Jones, P.J.H., 2009. Anticancer effects of phytosterols. *Eur. J. Clin. Nutr.* 63, 813–820. doi:10.1038/ejcn.2009.29

Yeganeh, B., Moshtaghi-Kashanian, G.-R., DeClercq, V., Moghadasian, M.H., 2005. Combination of dietary phytosterols plus niacin or fenofibrate: effects on lipid profile and atherosclerosis in apo E-KO mice. *The Journal of Nutritional Biochemistry* 16, 222–228. doi:10.1016/j.jnutbio. 2004.12.003

In: Phytosterols
Editor: Deanna Garner

ISBN: 978-1-63483-477-3


*Chapter 3*

# RECOVERY OF PHYTOSTEROLS FROM REFINING BY-PRODUCTS OF VEGETABLE OILS

***M. Victoria Ruiz-Méndez[1,*], Joaquín Velasco[1] and Gloria Márquez-Ruiz[2]***
[1]Instituto de la Grasa, CSIC
[2]Instituto de Ciencia y Tecnología de Alimentos y Nutrición, CSIC

## ABSTRACT

Refining of oils allows the elimination of color, odor and flavor that make crude oils unacceptable to consumers. In addition, chemical compounds that might be toxic or those that decrease the oil stability are removed in the refining process. Unfortunately, other components including phytosterols, of well-known bioactivity, are removed in part along with the undesired compounds and are ultimately concentrated in the refining by-products. Therefore, re-use and exploitation of by-products as an attractive source of bioactive compounds is gaining much interest. In this chapter, by-products from both chemical and physical refining used as sources of phytosterols, namely soapstocks and deodorization distillates, will be discussed. Soapstocks consist mainly of sodium salts produced from free fatty acids and also contain neutral oil and gums. Deodorization distillates include a considerable number of bioactive compounds, among them phytosterols, that make them excellent starting materials for obtaining novel food products of high added value. The extraction, isolation and purification of phytosterols from these raw materials using diverse technologies, including molecular distillation, extraction with supercritical carbon dioxide, hydrolysis of esters and crystallization and re-crystallization of adducts, will be revised.

* Corresponding author e-mail: mvruiz@ig.csic.es

# 1. Introduction

Phytosterols or plant sterols are substances of lipidic nature characteristic of oleaginous plants, nuts, seeds and cereals, where they act as important structural components of cell membranes. They are triterpenes with a chemical structure similar to that of cholesterol. Their basic structure is formed by a tetracyclic ring of cyclopentane-perhydrophenantrene to which a hydrocarbon chain with 8-10 carbon atoms is joined at C-17 position. Phytosterols include a hydroxyl group at C-3 position and most possess a double bound at C-5 (delta 5-sterols) or at C-7 (delta 7-sterols) positions. In addition, they can bear a second double bound in the hydrocarbon chain (Figure 1).

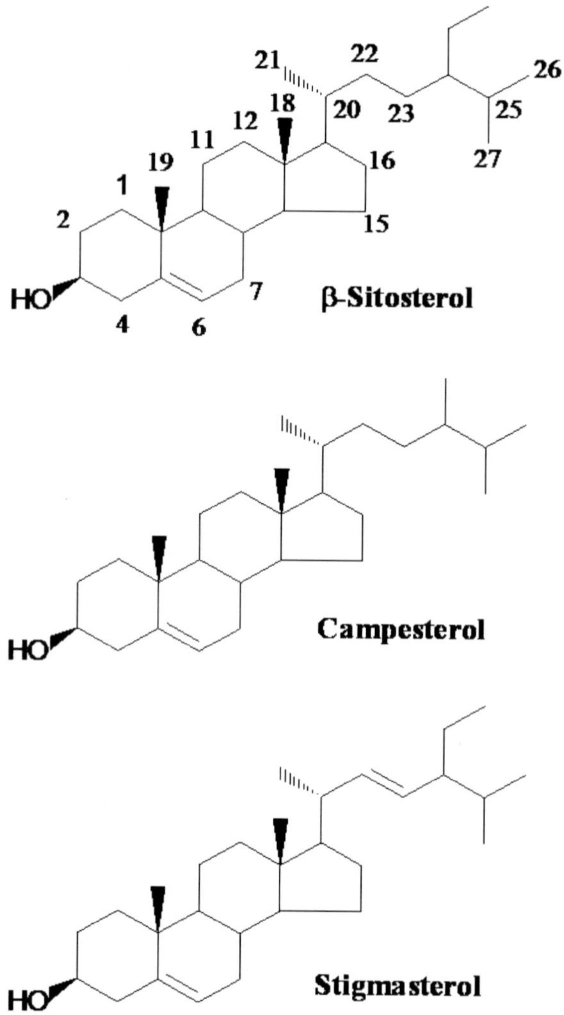


Figure 1. Chemical structures of the most abundant species of phytosterols.

More than 200 different types of phytosterols have been described and among them beta-sitosterol, campesterol and stigmasterol are the most abundant compounds (Piironen et al. 2004). Their structures differ from that of cholesterol mainly in the presence of a methyl or ethyl group at C-24 position (Figure 1). Unlike cholesterol, the structure of vegetable sterols may also include methyl groups at C-4 position and thereby three subclasses of sterols can be distinguished, namely, 4,4-dimethyl sterols, containing two methyl groups, 4-monomethylsterols, containing one methyl group, and 4-desmethylsterols, which do not possess additional methyl groups (Herchi et al. 2009). The first two subclasses are intermediates in the synthesis of sterols, being 4-desmethylsterols those that occur in plants in significant amounts. Phytosterols are present in nature both in free and conjugated forms. The hydroxyl group can be found either esterified with a fatty acid thus resulting in sterol esters, or glycosylated with a hexose thus constituting glucoside sterols. These latter, in turn, can be esterified with a fatty acid at C-6 position of the hexose moiety resulting in acylated glucoside sterols (Piironen et al. 2000).

Many positive physiological effects, such as anti-inflammatory, antitumor, bactericide and antifungal effects, have been attributed to phytosterols. Nevertheless, their best known health-promoting property is their hypocholesterolemic capacity. Inclusion of plant sterols and/or their saturated homologues stanols in the diet has been reported to lower concentrations of serum cholesterol associated to low-density lipoproteins up to 10% (Lauer & Fontanarosa, 2001; Katan et al. 2003).

Although such an effect has been known for over 60 years (Best et al. 1953; Pollak, 1953a; b), the mechanism of action still remains to be elucidated. The early mechanisms proposed suggested that plant sterols and/or stanols compete with intestinal cholesterol for their incorporation into mixed micelles prior to absorption and later also into chylomicrons. Then, main attention was focused on the role of sterol transporters located in the membranes of enterocytes. Plant sterols have been suggested to interact with intracellular cholesterol sensors that regulate the transporters responsible for the promotion of the efflux of sterols from enterocytes back into the intestinal lumen (De Smet et al. 2012). Recently, it has been suggested that plant sterols and/or stanols may activate directly secretion of cholesterol from the circulation into the intestinal lumen through a process called transintestinal cholesterol excretion, accounting for the increased fecal neutral cholesterol content and cholesterol-lowering activity observed (De Smet et al. 2012).

It has been estimated that the daily intake of plant sterols from conventional foods in different populations is between 160 and 400 mg (Berger & Jones, 2004). Such figures are based on the consumption of vegetable oils, nuts, seeds and grains. Vegetable oils, such as corn, sunflower, soybean and rapeseed oils, with contents between 0.1 and 0.9%, constitute the main dietary source of sterols (Piironen et al. 2000). For example, 30 g corn oil contain approximately 300 mg vegetable sterols, an amount high enough to reduce significantly cholesterol absorption (Ostlund et al. 2002a; b).

Supplementation of margarines with sterols and/or stanols was introduced in the European market more than 15 years ago. Results showed that 2 g of plant sterols or stanols added to an average daily portion of margarine reduced serum concentrations of low-density-lipoprotein cholesterol by an average of 0.54 mmol/L in people aged 50-59, 0.43 mmol/L in those aged 40-49 and 0.33 mmol/L in those aged 30-39. From such reductions, a decrease in the risk of heart disease of about 25% would be expected, which is larger than the effect expected to be achieved by people reducing their intake of saturated fat (Law, 2000).

Vegetable oils and, more specifically, by-products generated in the refining of oils, constitute the main sources of phytosterols for food supplementation purposes. Plant sterols, tocopherols, and other minor unsaponifiable components are removed from refined oils during deodorization and can be recovered from the deodorizer sludge. The major refined vegetable sources of plant sterols include soybean oil, corn oil, sunflower oil, rapeseed oil and wheat germ oil. Another relevant source of plant sterols is tall oil, a by-product of the paper pulping industry mainly produced from *Pinus spp* (Kamal-Edin & Moazzami, 2009).

In addition, minor amounts of phytosterols can also be found in the soapstocks obtained in the neutralization step of classical refining. Their levels depend on the initial contents in the crude oil and the refining conditions. This chapter will focus on the recovery of phytosterols from soapstocks and deodorization distillates. Given the significance of the refining and processing conditions on the phytosterol levels in these by-products, the refining process of oils will be first introduced.

## 2. The Refining Process of Oils

Vegetable oils are obtained by pressing and/or solvent extraction of oleaginous seeds and fruit.

The oils thus obtained, directly from extraction, are called crude oils. These mainly contain triacylglycerols and varying levels of non-triacylglycerol materials, which are mostly considered impurities in the finished oils.

However, the vast majority of oils which are sold to food processors and at a retail level are fully refined.

The refining process is very similar for all oils although some parameters may vary slightly according to the type of seed or source. Overall, the ultimate goal is always to yield a clean and stable product with 96% or more triacylglycerols.

Refining of crude oils is applied to remove unwanted minor components that make oils unappealing to consumers, while reducing oil damage and loss in so far as possible.

The components to be removed are all those glyceridic and non-glyceridic compounds that negatively affect flavor, color, stability or safety of the refined oil. They are primarily phospholipids, free fatty acids (FFA), pigments, volatiles and contaminants. However, not all the minor compounds in fats and oils are undesirable. For example, phytosterols and antioxidants that protect the oil from oxidation such as tocopherols with vitamin E activity are highly appreciated. Consequently, in order to reach maximum oil quality, all the steps of the refining process should be carried out with minimum losses of desirable compounds.

The major steps of oil refining and the main components removed are shown in Table 1. Two standard processes, alkali or chemical and physical refining, are those applied in the fats and oils industry.

The main difference between alkali and physical refining is that in the former a caustic soda treatment is applied to neutralize the oil, while in the latter FFA are eliminated by distillation during deodorization. Physical refining reduces the loss of neutral oil, minimizes pollution and enables recovery of high quality FFA. Nevertheless, not all oils can be physically refined.

**Table 1. Basic steps of the refining process of oils**

| Alkali or chemical refining | Main groups of compounds removed | Physical Refining |
|---|---|---|
| Degumming | Phospholipids | Degumming |
| Neutralization | Free fatty acids | - |
| Bleaching | Pigments/metals/soaps | Bleaching |
| Winterization | Waxes/saturated triacylglycerols | Winterization |
| Deodorization | Volatiles/Free fatty acids | Deodorization/ Deacidification |

## Degumming

The purpose of degumming is to remove phospholipids or gums from the crude oil. Two types of phospholipids are present according to their level of hydration, i.e., hydratable and non-hydratable, the latter mainly comprising calcium and magnesium salts of phosphatidic and lysophosphatidic acids. After addition of water (1-3%), most of the phospholipids are hydrated and become insoluble in the oil. Hydrated phospholipids can be efficiently separated by filtration or centrifugation. To eliminate the non-hydratable fraction, the oil is usually treated with phosphoric acid (0.05 to 1%), which dissociates the phosphatidates converting them into the hydratable forms and chelates the Ca and Mg ions.

Depending on the oil composition, the degumming step can be omitted when the phosphatidates are removed along with the soaps during the next step of neutralization. However, degumming is mandatory for physical refining and the content of phosphorous after degumming must be lower than 10 mg/kg (Deffense, 2009, De Greyt, 2013).

## Neutralization

In this step, the oil is treated with caustic soda (sodium hydroxide) and hence the FFA are converted into insoluble soaps which can be easily separated by centrifugation. Although the main objective of neutralization is the removal of FFA, residual phospholipids in degummed oils or all the phospholipids in crude oils are concomitantly removed as insoluble hydrates. In addition, caustic neutralization also contributes to partial bleaching since polar pigments, such as gossypol, sesamol, sterols, hydroxy fatty acids and others, react in part with sodium hydroxide and then are also removed by water solubilization. Alkali refining is absolutely necessary in crude oils of high acidity and/or high pigment contents.

The free fatty acid content of the oil is the main factor that determines the amount and concentration of the caustic soda used and also its excess (5 to 20%) for a minimum oil loss. After a reaction time of about 30 min, at slow stirring and temperature around 80 °C, the water phase containing the fatty soaps is eliminated by centrifugation and the oil is washed with water to remove residual soaps. The by-product generated in the process is termed soapstock and contains essentially 10-60% water, 40-90% fatty acid soaps and minor contents of neutral oil.

During alkali neutralization, a reduction in the total content of sterols also occurs. Such a reduction has been attributed mainly to losses of free sterols that may form micelles with soaps and then be transferred into the soapstock (Verleyen et al. 2002). Contents up to 21% sterols from the crude oil can be thus transferred into soapstocks (Ubhayasekera & Dutta, 2009).

## Bleaching

In this step, which is common to physical and alkali refining processes, the oil is heated at about 100ºC and treated with acid-activated bleaching earth (1-2%), normally calcium montmorillonite or natural hydrated aluminum silicate (bentonite). Under these conditions, adsorption of color bodies, trace metals and oxidation products as well as residual soaps and remaining phospholipids takes place. For optimum adsorption of both color bodies and oxidation products, the reaction time has to exceed 15 min but should not surpass 30 min at usual bleaching temperatures. The removal of chlorophyllic pigments is essential since, unlike carotenoid compounds that are removed during deodorization, they are not eliminated in any other stage of refining. Finally, filtration must eliminate completely the activated earths as the presence of traces would act as prooxidants during oil storage because of their iron content.

Acid-activated clays are the major adsorbents used, although active carbons and synthetic silicas are also applied industrially with more specific goals. Thus, active carbons are used specifically to eliminate polycyclic aromatic hydrocarbons (PAH) from some oils, especially fish oils and pomace oils, whereas synthetic silicas are otherwise selected for their efficiency in adsorbing secondary oxidation products, phospholipids and soaps (Hussin et al. 2011).

Alcohols or hydroxy acids formed from hydroperoxides undergo a partial dehydration by earth catalysis resulting in a new carbon-carbon double bond. As the hydroxy group is normally at an allylic position, a rapid increase in UV absorption at 232 nm is observed because of the formation of conjugated dienes from oleic acid hydroxides, and in UV absorption at 270 nm due to formation of conjugated trienes from linoleic acid hydroxides. Also, sterols undergo significant dehydration and the formation of the hydrocarbon 3,5-stigmastadiene from the major sterol (β-sitosterol) is considered a proof of the presence of refined oil in virgin olive oil (Lanzón et al., 1994).

## Winterization

The winterization or dewaxing step is only applied when the oil is not clear at room temperature because of the presence of waxes or saturated triacylglycerols. Such compounds do not affect negatively the oil performance or functionality, but they change the oil appearance, making it unacceptable to consumers.

The objective of this step is therefore the removal of high temperature melting components present in small quantities. The crystallization process normally used consists in cooling the oil down gradually to temperatures about 5-8 ºC in a maturing tank. After increasing the crystal size at 5-8 ºC from 24 to 48 h, the solids are separated by centrifugation at 15-16 ºC. This treatment guarantees excellent clarity of oils when stored at either room or refrigeration temperatures.

### Deodorization/De-Acidification

Deodorization of fats and oils normally consists in steam distillation at elevated temperature under reduced pressure, although nitrogen has also been used (Ruiz-Méndez et al. 1996). The purpose of this step is to remove volatile compounds, mainly ketones and aldehydes that affect oil taste and odor, total FFA in the case of physical refining and the residual FFA in the case of neutralized bleached oils. The deodorization conditions also favor the removal of contaminants, including light PAH and pesticides, and improve color reduction as a result of the breakdown of remaining carotenes at high temperature. The efficiency of deodorization is a function of pressure (1 to 5 torr), temperature (200 to 260 ºC), residence time (0.5 to 3 h) and volume of stripping gas (1 to 3%).

Long deodorization times and/or too high temperatures can have a devastating effect on the oil quality, not only due to the chemical changes commented above, but also to the distillation of a significant part of the natural free sterols and tocopherols (20 to 40%), whose elimination would decrease the stability of the refined oil (Ergönül & Köseoğlu, 2014).

Modern commercial deodorizers are equipped with a packed column with three sections: vapor scrubbing, stripping and heat bleaching sections. Prior to entering the vacuum equipment, the vapors pass to a condenser (scrubber). A scrubbing liquid is sprayed in the vapor stream and the condensates are known as "deodorizer distillates (DOD)", an important by-product. It represents between 0.1 and 0.4% of the oil to be refined, and its composition varies according to the type of oil and the deodorization conditions (Winters, 1990). Most deodorizers have only one vapor condenser from which one single deodorizer sludge is collected. The vapor stream from the deodorizer comes together with a liquid stream obtained by condensation of the very distillate, i.e., the distillate is recirculated to join with the deodorizer vapor. It is necessary to control the temperature of the liquid stream at a temperature at which the whole distillate condenses (Jackson & Reynolds, 1968).

Due to the growing interest in and demand for tocopherol-rich distillates (as a source of natural vitamin E) the current trend is to improve the design of the vapor scrubber system aiming at a further reduction of neutral oil losses produced by mechanical entrainment of small oil drops in the steam stream (De Greyt, 2013). A "dual condensation" system, so-called 'double scrubber', with high and low temperatures condensation zones, has been developed (Dijkstra, 2013). At the high- temperature condensation zone, compounds of high added value are quickly cooled and quantitatively captured in the cold condensation zone. Thus, an essentially no-loss operation for the micronutrients is achieved, leading to an enriched micronutrient stream with 20.0 wt % tocopherol and 31.4 wt% sterols (Sarup & Danilo, 2014).

## 3. Recovery of Phytosterols from Soapstocks

The main interest of vegetable oil soapstock reuse is obtaining fatty acids. Isolation of fatty acids from soapstocks has been traditionally driven primarily towards animal feed and to a much lesser extent towards oleochemical applications. Low-grade cleaning, soap and other surfactant products are manufactured from vegetable oil soapstocks. In this context, there are a few comprehensive reviews dealing with isolation of fatty acids from vegetable oil soapstocks (Woerfel, 1983; Sonntag, 1990; Dumont & Narine, 2007). The main process applied in the industry is the one termed soapstock splitting, on account of easiness and low cost. It consists in soapstock acidulation with sulphuric acid followed by centrifugal separation.

Although it cannot compete with DOD, soapstock is also a source of phytosterols. It normally consists of 10-60% water, and the rest is constituted by fatty acid soaps, 0.1-2% sterols, mono-, di- and triacylglycerols and low levels of phospholipids. In some cases there are also proteins that end up as solid material in the process (Kempers et al. 2013). As it will be discussed later, DOD availability is much greater as compared to that of soapstocks and hence DOD constitute a more suitable source of phytosterols and other oil minor components with added value, including tocopherols, squalene and others.

Reports on the utilization of soapstocks to obtain phytosterols are rather scant. In this respect, studies have been mainly focused on oryzanol, a group of sterol derivatives named ferulic acid esters of sterols or steryl ferulates, of nutritional and commercial interest. Steryl ferulates have been identified in rice, corn, wheat, rye, barley and others, exhibiting rice (*Oryza sativa L.*) the highest levels. The group of steryl ferulates found in rice is normally termed γ-oryzanol (Figure 2). Rice bran is a valuable by-product of the rice processing industry and extraction of oil from rice bran is in many cases economically feasible.

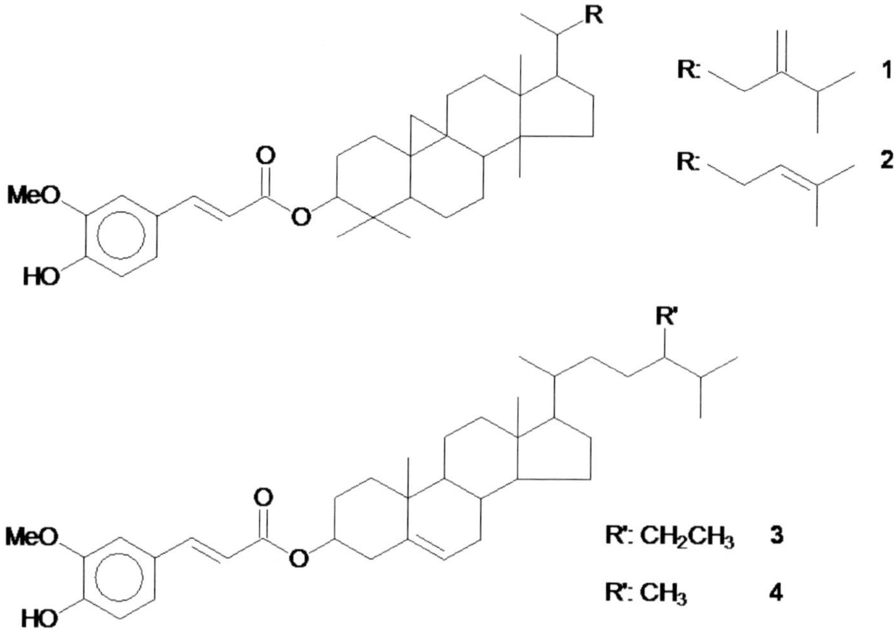


Figure 2. Chemical structures of main components of γ-oryzanol. (1) 24-Methylen-cycloartanylferulate. (2) Cycloartanylferulate. (3) β-Sitosterylferulate. (4) Campesterylferulate.

However, refining of crude rice bran oil (RBO) presents difficulties due to the presence of high levels of FFA, waxes, gums and pigments. The soapstocks generated in chemical refining contain approximately 65-70 wt% water, 20-22 wt% soap, 2-2.5 wt% glycerides, mainly triacylglycerols, and 7-7.5 wt% unsaponifiable material (Narayan et al. 2006). It has been reported that the unsaponifiable matter consists of approximately 42 wt% sterols, 24 wt% fatty alcohols, 20 wt% oryzanol, as ferulic acid esters, 10% hydrocarbons and 2 wt% unidentified compounds (Akiya, 1962). Therefore, soapstock from the chemical refining of RBO is the major source of oryzanol.

Numerous patented approaches for recovery and purification of oryzanol from RBO soapstocks (RBOS) include solid-liquid extraction, liquid-liquid extraction, supercritical fluid extraction, crystallization or precipitation and combinations thereof (Narayan et al. 2006). A few examples will be described next.

Solid-liquid extraction has been applied for both isolation of oryzanol from dried RBOS and removal of impurities from oryzanol-enriched fractions. Oryzanol has been obtained with 85% purity from a procedure of two steps (Masao & Yoshizane, 1968). First, soaps were selectively removed with methanol or ethanol and simultaneous purging with carbon dioxide gas for acidification of the medium. Then weak alkaline salts such as sodium bicarbonate and sodium carbonate, associated with the residue of the first extraction step, were removed with methanol or ethanol. Another procedure consisted in the treatment of RBOS with sulphuric acid followed by esterification of FFA with methanol and subsequent distillation of fatty acids as fatty acid methyl esters. The residue remaining after distillation was subjected to solid-liquid extraction with n-hexane, methanol and methanolic alkali (Takeshi, 1969). In another approach, RBOS was previously saponified and dehydrated. Then extractions with ethyl acetate, acetone or mixtures thereof were applied to obtain oryzanol of 40-45 wt% purity and at 80 wt% yield (Indira et al. 2005).

Oryzanol of 98 wt% purity has been obtained by multiphase fractional crystallization from the soapstock derived from a second alkali-refining process of RBO (Mingzhi & Yanyan, 1997). Starting from the unsaponifiable material obtained from RBOS by saponification of the neutral oil, dehydration and solid-liquid extraction, Rao et al. (2002) isolated oryzanol with 90 wt% purity and at 56-70 wt% yield by crystallization, column chromatography and recrystallization. However, the use of column chromatography makes it difficult to scale up such a procedure for industrial production purposes. Similarly, Narayan et al. (2005) obtained oryzanol with 65 wt% purity and at 70 wt% yield from the unsaponifiable material of RBOS by crystallization. The unsaponifiable material was obtained in a similar way to that of Rao et al. (2002). The crystallization was performed with combinations of acetone and methanol at reflux conditions followed by cooling at room temperature. Mucilaginous impurities, i.e., waxes, precipitated and were removed and the system was further cooled to 5-10 °C overnight for oryzanol crystallization.

Other procedures reported for oryzanol recovery consisted in a four-step process that includes liquid-liquid extraction, column chromatography, crystallization and recrystallization (Seetharamiah & Prabhakar, 1986).

Recovery of sterol esters from soapstocks has been also approached by enzymatic esterification of FFA with short chain alcohols (C1 to C6), preferably methanol or ethanol, followed by distillation (Kempers et al. 2013). The soapstocks can be previously treated with alkaline lipase to hydrolyze the glyceridic components or be subjected to neutralization with strong acids followed by separation of soaps and subsequent enzymatic hydrolysis of glycerides. Then FFA are selectively esterified with lipases, while sterol esters remain intact during the treatment. The fatty acid alkyl esters produced are removed by distillation, preferably by short path distillation, and the sterol esters are enriched in the heavy or bottom fraction. The enzymatic esterification of FFA was compared to chemical esterification with zinc oxide. The results showed that the total sterol content in the distillation residue of the chemically treated sample was 8.8% being 40% sterol esters, while the total content of sterols in the enzymatically treated sample was 4.6% being 79% sterol esters. Therefore, whereas the sterol esters do not react to a significant extent during the enzymatic esterification of the FFA, a significant part of them is otherwise transformed during the chemically catalyzed esterification.

## 4. Recovery of Phytosterols from Deodorization Distillates

Only small amounts of sterols are removed in the neutralization and bleaching steps while the main sterol removal occurs during deodorization. This is due to the relatively high vapor pressure of sterols at elevated temperatures and high vacuum (Antonopoulos et al. 2006).

Composition of DOD varies greatly depending on the starting oil, the type of refining process and the operation conditions applied. Table 2 shows analytical data for DOD produced in physical and chemical refining of different vegetable oils using a discontinuous device. The results correspond to main groups of compounds analyzed by thin-layer chromatography coupled with flame ionization detector (TLC-FID) (Ruiz-Méndez et al. 1995). Five groups of compounds can be determined in the TLC-FID analysis, namely, FFA, triacylglycerols, sterols, hydrocarbons and other minor components. The data showed marked differences between the distillates obtained in the two processes.

Triacylglycerols coming from mechanical entrainment were the major compounds present in the DOD of chemical refining. In contrast, FFA were the most abundant components in the DOD of physical refining, thereby the fatty-acids/triacylglycerols ratio was much higher.

Distillates from the physical refining of oils, such as palm oil and olive oil, have typically acidity values above 70% and are mainly used in the preparation of fatty acids or biodiesel (Echim et al. 2009). Olive oil DOD are also appreciated by its high content in squalene (Bondioli et al. 1993). In contrast, DOD from chemical refining are characterized by a high content of unsaponifiable fraction (15-30%) since FFA of the crude oil are previously removed in the neutralization step. Pomace olive oil DOD represent a special case, in which alkyl esters constitute the major components. The fermentation process that occurs during olive-pomace storage in ponds yields esters of fatty acids with linear short chain alcohols (alkyl esters). They are completely lost during deodorization, the only step in the refining process in which their elimination is foreseen (Ruiz-Méndez & Dobarganes, 2007).

Of great significance is the elevated concentrations of pesticides present in DOD, which can be found at levels as much as 800–1000 times higher than the values allowed in foods (Ferrario et al. 2002; Halber & Atcher, 2007).

**Table 2. Composition of deodorizer distillates produced in the physical and chemical refining of different vegetable oils as determined by thin-layer chromatography coupled with flame ionization detector (TLC-FID)**

| | Physical Refining* | | Chemical Refining** | | | |
|---|---|---|---|---|---|---|
| | 1 | 2 | 3 | 4 | 5 | 6 |
| Free fatty acids (%) | 70.7 | 63.6 | 35.0 | 25.3 | 34.2 | 29.6 |
| Triacylglycerols (%) | 4.1 | 10.7 | 42.7 | 48.0 | 40.7 | 44.3 |
| Sterols (%) | 0.6 | 0.6 | 4.1 | 2.3 | 6.7 | 6.3 |
| Hydrocarbons (%) | 15.3 | 14.2 | 5.2 | 10.3 | 5.8 | 4.1 |
| Others (%) | 9.3 | 10.9 | 13.0 | 14.1 | 12.6 | 15.7 |

*Distillates obtained from olive oils. **Distillates obtained from different seed oils.

Therefore, the valorization of DOD as food additives rich in bioactive compounds markedly depends on the effective removal of pesticides, which is nowadays a technological challenge (Teixeira et al. 2014).

Preparing pure sterol compounds from DOD, where they are present in low concentrations as secondary substances, implies a complex process. The components of higher price are tocopherols (Vitamin E), which are natural antioxidants. Sterols are also of high value but this depends on the type of oil (Laoretani & Iribarren, 2014). Excellent reviews have been published on separation strategies and technologies to obtain bioactive compounds from DOD (Fernandes & Cabral, 2007; Gunawuan & Ju, 2009; Kamal-Eldin & Moazzami, 2009; Torres et al. 2011; Yan et al. 2012; Carmona et al. 2015). In fact, a large number of procedures to extract and purify plant sterol/stanol esters from DOD have been patented (Kamal-Eldin & Moazzami, 2009). We can differentiate them into two main groups, namely, those aimed at obtaining unsaponifiable concentrates and those which, starting from the concentrates or directly from DOD, are directed to isolate phytosterols.

## Unsaponifiable Concentrates

As remarked above, the three main constituents of DOD are FFA, acylglycerols and unsaponifiable fraction. Unsaponifiable concentrates can be obtained by hydrolysis of glycerides to FFA, saponification of FFA and glycerides into soaps, and esterification of FFA (Laoretani & Iribarren, 2014). Hydrolysis can be carried out at strong basic conditions (saponification), which requires a further acidulation, or at strong acidic conditions, by chemical or enzymatic catalysis (Fernandes & Cabral, 2007; Yan et al., 2012; Benites et al. 2014).

Given that part of phytosterols is present in raw materials esterified with fatty acids (Gunawuan & Ju, 2009), a pre-treatment can be applied to release simple phytosterols. An approach that does not require a large investment is chemical hydrolysis. DOD are hydrolyzed in an autoclave under high pressure and temperature (1.5-50 MPa and 200-260°C). The process is then followed by distillation of the fatty acids to concentrate tocopherols and phytosterols in the residue (Daguet & Coïc, 1999). Nevertheless, thermal degradation can occur if the distillate is exposed for extended periods to the temperatures at which phytosterols vaporize (Copeland & Belcher, 2003). To remove FFA some authors have proposed further distillation of DOD by using molecular distillation (Martin et al. 2006; Sawadikiat & Hongsprabhas, 2014). In order to optimize the process, Pramparo et al. (2005) carried out separation tests by molecular distillation with and without previous esterification of fatty acids. Esterification gave higher separation efficiency in the first stage of molecular distillation in all the tests. The second stage of molecular distillation allowed increasing concentrations of tocopherols and sterols. Esterification of fatty acids or phytosterol fatty acid esters can be catalyzed by metal alcoholates or metal hydroxides like zinc oxide or zinc hydroxide, by alkali metal alcoholates like sodium methoxide and by organic catalysts, like monoalkyltin or dialkyltin compounds or phenyl phosphonic acid (Hoydonckx, et al. 2004).

The most used alcohol is methanol because it leads to methyl esters, which show high volatility, but other monohydric C2 to C4 alcohols can be alternatively used. This latter approach, using methanol, is the one that arouses most interest due to the high price that by-products can reach (Echim et al. 2009). Gole and Gogate (2012) proposed an intensification of the synthesis of fatty acid methyl esters (FAME) using *sonochemical* reactors. More recently, Yin et al. (2015) have reported the use of *sonochemical* reactions as a technique for intensification of biodiesel synthesis from soybean oil DOD at static or counter-current way. ''Counter-current'' means that the direction of the ultrasound wave is from up to down, while the flow direction of the reactor is the opposite. This kind of working mode allowed efficient contact between the ultrasound and the reactants.

The use of heterogeneous catalysts could be an interesting approach for FAME production because catalysts can be separated more easily from reaction products and undesired saponification reactions can be avoided (Lee et al. 2014). Nevertheless, today, the novel interest for using natural organic molecules for the production of natural grade chemicals, for the hemisynthesis of complex molecules or for the sustainable use of renewable resources is also observed in enzymatic catalysis in organic synthesis (Groussin & Antoniotti, 2012). In this context, microbial lipase-catalyzed methyl or ethyl esterification of DOD from soybean oil followed by molecular distillation and/or supercritical fluid extraction has been proposed (Panpipat, et al. 2012).

The most remarkable difference between chemical and enzymatic reactions is the fact that lipase possesses several substrate specificities, in particular, fatty acid specificity on hydrolysis, esterification or transesterification reactions of the ester bond. Lipase specificity also facilitates purification of PUFAs, conjugated linoleic acid isomers, tocopherols, sterol esters and triacylglycerols rich in DHA at the *sn*-1,3 positions (Nagao et al. 2013).

## Isolation of Phytosterols

Recovery of the phytosterol fraction can be approached from unsaponifiable concentrates using three types of extraction procedures, namely, physical extraction, by which phytosterols are crystallized and physically separated, chemical extraction with solvents, and physicochemical extraction, by which phytosterols are crystallized as adducts with additives (Daguet & Coïc, 1999).

Crystallization of phytosterols can be obtained by solvents (Lin & Koseoglu, 2003; Moreira & Baltanás, 2004; Carmona et al. 2010; Yang et al. 2010). The phytosterols crystallize or otherwise precipitate to form a solid phase that can be physically separated from the liquid phase, e.g., by filtration, centrifugation or settling. Lin and Koseoglu (2003) achieved their best results without pre-treatment by crystallization at -20 °C for 24 h with a mixture of acetone and methanol, followed by centrifugation, filtration and washing. Over 90% of the original tocopherols and squalene in DOD were recovered in the filtrated fraction, while 80% of total sterols in DOD were retained in the cake fraction.

Pramparo et al. (2005) tested crystallization of the sunflower distillated final stream by molecular distillation of an unsaponifiable concentrate, thus allowing separation of more than 40% of sterols and obtaining a product with more than 92% w/w of sterols.

Chemical and physical extractions can be combined so that phytosterols can form adducts with additives that crystallize and then can be physically recovered. Subsequent isolations of phytosterols from esterified and transesterified mixtures by adduct formation can be achieved by using an aprotic solvent containing a small amount of a protic solvent.

After separation, the adducts are decomposed in a biphasic solvent system to release free sterols and additives.

The phytosterol fraction can also be extracted by successive liquid/liquid extractions with combinations of polar and non-polar solvents. The solvent blend should comprise a major amount of a low polarity solvent, typically at least about 80 wt%, and a solvent of high polarity, normally at levels less than 20 wt%. Water is also used in the blend at a high-polarity-solvent-to-water ratio that ranges between 5:1 and 1:5 (Daguet & Coïc, 1999).

Another approach tested has been formation of steryl esters by enzymatic esterification of free fatty acids and sterols present in DOD (Torres et al. 2011; Panpipat, et al. 2012; Teixeira et al. 2012).

The enzyme selected is normally a *Candida rugosa* lipase, which has shown to be able to synthesize steryl esters at a yield of 80% at optimized conditions for 24 h. Teixeira et al. (2014) applied and optimized this strategy in order to make it easier the recovery of phytosterol by membrane technology.

In spite of the considerable number of technologies proposed for the separation between tocopherols, sterols and FFA (Fernandes & Cabral, 2007), molecular distillation first and supercritical extraction with $CO_2$ afterwards have been the ones that found their way into industrial scale production (Laoretani & Iribarren, 2014).

Molecular distillation devices are commonly used to prevent thermal degradation of liquid material such as high-molecular weight organic liquids, whose components have a low vapor pressure at conventional distillation temperatures. In such devices the distance between the hot and cold surfaces is typically about the average mean free path of the molecular fraction to be distilled.

Phytosterols cannot be removed independently by molecular distillation because their molecular weights and volatilities are similar to those of tocopherols (Ghosh, & Bhattacharyya, 1996). Concentration of phytosterols in good yield and high quality is a difficult task. Esterification of sterols with the fatty acids already present in the DOD has been proposed to make the separation of tocopherols and sterols easier using short-path distillation and/or supercritical fluid.

Isolation of phytosterols from DOD can be also approached by supercritical fluid extraction with $CO_2$ (SC-$CO_2$). The extraction of phytosterols is much faster than that of the oil (triglycerides) and thus phytosterols can be concentrated in the extract using low solvent volumes. High pressure and moderate temperatures favor phytosterol extraction from various plant matrices. A few research studies have been performed to isolate phytosterol esters from soybean oil DOD by counter-current extraction using SC-$CO_2$ (Brunner, 2005; Uddin et al. 2015). In counter-current extraction, $CO_2$ flows continuously from the bottom of a column and the liquid sample is pumped from the top. Before extraction, the raw material is first subjected to enzymatic reactions to obtain a mixture consisting of fatty acid ethyl esters, tocopherols and phytosterol esters, with minor amounts of free fatty acids, squalene, free sterols and triacylglycerols.

Phytosterol esters are then separated from this mixture by countercurrent extraction using SC-$CO_2$. The extraction is carried out at pressures of 20-28 MPa, temperatures of 45-55.7 ºC and solvent-to-feed ratios of 15-35 kg $kg^{-1}$.

Membrane-based separation constitutes an emerging technological process for phytosterols recovery. Teixeira et al. (2014) studied solvent resistant nanofiltration (SRNF) as a potential alternative technology for recovery of bioactive compounds, being remarkably simpler than traditional processes such as distillation, solvent extraction and crystallization, which are labor-intensive, time consuming and energetically intensive. On one hand, SRNF enabled the removal of 99.2% of the pesticides from esterified-deodorizer distillate, below 0.05 ppm in the enriched-food, which is the limit value to be complied. Such an efficient pesticide removal was achieved, however, at expense of a significant loss of steryl esters (42%).

On the other, SRNF enabled purification, fractionation and concentration of natural extracts without requiring high processing temperatures which typically damage their valuable compounds. In order to narrow the large number of alternatives published for recovery of phytosterols, Laoretani and Iribarren (2014) have suggested an interesting approach.

They focused on how to generate a first set of promising alternatives among the current strategies by following an established process-design procedure, based on heuristics, combined with a screening of the literature. The final selection is an approach that implements a multi-objective optimization program maximizing the net present value (NPV) and minimizing the generation of greenhouse gases measured as kilogram-equivalent of $CO_2$. For a given case study of soybean DDO, the first step generated a set of six technologies for the treatment of DDO with different processing capacities plus two additional alternatives for the final destination of DDO. A Pareto set of solutions constructed with the results provides information to adopt an economically and environmentally sound choice of a processing technology. For the particular case analyzed, the technology that maximized NPV within the Pareto set of solutions was esterification of free fatty acids with ethanol in acid medium followed by separation of the esters by molecular distillation, at the largest production capacity (576,000 kg/year). This technology gives the maximum NPV of $19,574,000 generating 5,142,500 kg of $CO_2$-equiv. The results obtained are useful in the industry to achieve an adequate final destination for DDO residues.

## Conclusion

In spite of the numerous processes that have been proposed for recovering phytosterols from refining by-products, efficiency and economically feasible technologies still continue being a challenge. New raw materials for the recovery of bioactive compounds, including phytosterols, such as the by-products generated in the industrial production of biodiesel, will be considered in the short term. Such new potential substrates will require adaptation of the procedures applied at present or the development of new sustainable technologies.

## ACKNOWLEDGMENTS

This work was financially supported by "Ministerio de Economía y Competitividad" through project AGL2013-45110-R, "Junta de Andalucía" through project P12-AGR-1837 and the European Community through project ERA-Net "SUSFOOD" Ref 291766.

## REFERENCES

Akiya, T. (1962) Components of the unsaponified matter of the rice bran oil. *Agric Biol Chem,* 26, 180-186.

Antonopoulos, K., Valet, N., Spiratos, D., & Siragakis, G. (2006). Olive oil and pomace olive oil processing. *Grasas y aceites*, *57*(1), 56-67.

Benites, C. I., Klein, B. C., & Reis, S. M. P. M. (2014). Neutralization of Soybean Oil Deodorizer Distillate for Vitamin Supplement Production. *International Journal of Chemical Engineering*, 2014.

Berger, A., Jones, P. J. H., & Abumweis, S. S. (2004). Plant sterols: factors affecting their efficacy and safety as functional food ingredients. *Lipids Health Dis, 3(5)*, 1-19.

Best, M. M., Duncan, C. H., Van Loon, E. J., & Wathen, J. D. (1954). Lowering of serum cholesterol by the administration of a plant sterol. *Circulation, 10(8)*, 201-206.

Bondioli, P., Mariani, C., Lanzani, A., Fedeli, E., & Muller, A. (1993). Squalene recovery from olive oil deodorizer distillates. *Journal of the American Oil Chemists' Society, 70*(8), 763-766.

Brunner, G. (2005). Supercritical fluids: technology and application to food processing. *Journal of Food Engineering, 67*(1), 21-33.

Carmona, M. A., Jiménez, C., Jiménez-Sanchidrián, C., Peña, F., & Rafael Ruiz, J. (2010). Isolation of sterols from sunflower oil deodorizer distillate. *Journal of Food Engineering, 101*(2), 210-213.

Carmona, M.A., Jimenez-Sanchidrian, C., & Rafael Ruiz, J. (2015). Recent developments in phytosterol recovery from oil deodorizer distillate*s. Current Nutrition & Food Science, 11*(1), 4-10.

Copeland, D., & Belcher, M.W. (2003). Methods for treating deodorizer distillate. WO03080778.

Daguet, D., & Coïc, J. P. (1999). Phytosterol extraction: state of the art. OCL. *Oléagineux, corps gras, lipides*, *6*(1), 25-28.

De Greyt, W.F.J. (2013). Edible oil refining: Current and future technologies. In: Edible Oil Processing, pp.127-151 (W. Hamm, R.J. Hamilton and G.H. Calliauw (eds), John Wiley & Sons, Chichester).

Deffense, E. (2009). From organic chemistry to fat and oil chemistry. *Oléagineux, Corps gras, Lipides*, *16*(1), 14-24.

De Smet, E., Mensink, R. P., & Plat, J. (2012). Effects of plant sterols and stanols on intestinal cholesterol metabolism: suggested mechanisms from past to present. *Molecular Nutrition & Food Research, 56(7)*, 1058-1072.

Dijkstra, A. J. (2013). Vacuum Stripping. In *Edible Oil Processing from a Patent Perspective* (pp. 205-218). Springer US.

Dumont, M. J., & Narine, S. S. (2007). Soapstock and deodorizer distillates from North American vegetable oils: Review on their characterization, extraction and utilization. *Food Research International, 40*(8), 957-974.

Echim, C., Verhé, R., De Greyt, W., & Stevens, C. (2009). Production of biodiesel from side-stream refining products. *Energy & Environmental Science, 2*(11), 1131-1141.

Ergönül, P. G., & Köseoğlu, O. (2014). Changes in α-, β-, γ-and δ-tocopherol contents of mostly consumed vegetable oils during refining process. *CyTA-Journal of Food, 12*(2), 199-202.

Fernandes, P., & Cabral, J. M. S. (2007). Phytosterols: applications and recovery methods. *Bioresource technology, 98*(12), 2335-2350.

Ferrario, J., Lovell, R., Gardner, P., Lorber, M., Winters, D., & Byrne, C. (2002). Analysis Of animal- and plant-derived feed ingredients for dioxin-like compounds. *Organohalogen Compounds, 57*, 85.

Ghosh, S., & Bhattacharyya, D. K. (1996). Isolation of tocopherol and sterol concentrate from sunflower oil deodorizer distillate. *Journal of the American Oil Chemists' Society, 73*(10), 1271-1274.

Gole, V. L., & Gogate, P. R. (2012). Intensification of synthesis of biodiesel from nonedible oils using sonochemical reactors. *Industrial & Engineering Chemistry Research, 51*(37), 11866-11874.

Groussin, A. L., & Antoniotti, S. (2012). Valuable chemicals by the enzymatic modification of molecules of natural origin: Terpenoids, steroids, phenolics and related compounds. *Bioresource technology*, *115*, 237-243.

Gunawan, S., & Ju, Y. H. (2009). Vegetable oil deodorizer distillate: characterization, utilization and analysis. *Separation & Purification Reviews,* 38(3), 207-241.

Halbert, M. K., & Archer, J. C. (2007). Dioxin and furan contamination of deodorizer distillates and natural vitamin E supplements. *Journal of food composition and analysis, 20*(6), 506-514.

Herchi, W., Harrabi, S., Sebei, K., Rochut, S., Boukhchina, S., Pepe, C., & Kallel, H. (2009). Phytosterols accumulation in the seeds of Linum usitatissimum L. *Plant Physiology and Biochemistry, 47(10)*, 880-885.

Hoydonckx, H. E., De Vos, D. E., Chavan, S. A., & Jacobs, P. A. (2004). Esterification and transesterification of renewable chemicals. *Topics in Catalysis*, *27*(1-4), 83-96.

Hussin, F., Aroua, M. K., & Daud, W. M. A. W. (2011). Textural characteristics, surface chemistry and activation of bleaching earth: A review. *Chemical Engineering Journal, 170*(1), 90-106.

Indira, T. N., Narayan, A. V., Barhate, R. S., Raghavarao, K. S. M. S., Khatoon, S., Gopal, C., Appurao, A. G., & Prakash V. (2005). Process for the production of oryzanol enriched fraction from rice bran oil soapstock, U.S. Patent 6,896,911.

Jackson, D. H., & Reynolds, J. T. (1968). "Method for the recovery of deodorizer distillates." U.S. Patent No. 3,369,344.

Kamal-Eldin, A., & Moazzami, A. (2009). Plant sterols and stanols as cholesterol-lowering ingredients in functional foods. *Recent Patents on Food, Nutrition & Agriculture, 1*(1), 1-14.

Katan, M. B., Grundy, S. M., Jones, P., Law, M., Miettinen, T., Paoletti, R., & Participants, S. W. (2003). Efficacy and safety of plant stanols and sterols in the management of blood cholesterol levels. *Mayo Clinic Proceedings*, 78(8), 965-978.

Kempers, P., Schörken, U., Wolf, T., Sato, S., Bueno, W., Silva, P., & Araujo, A. S. (2013) Process for production of fatty acids, fatty acid esters and sterolesters from soapstock. US Patent 8,426,622 B2. Washington, DC: U.S. Patent and Trademark Office.

Laoretani, D. S., & Iribarren, O. A. (2014). Procedure for the Selection among Technologies. Treatment of Deodorizer Distillate Oil. *Industrial & Engineering Chemistry Research, 53*(43), 16803-16812.

Lanzón, A., Albi, T., Cert, A., & Gracián, J. (1994). The hydrocarbon fraction of virgin olive oil and changes resulting from refining. *Journal of the American Oil Chemists' Society, 71*(2)285-291.

Lauer, M. S., & Fontanarosa, P. B. (2001). Updated guidelines for cholesterol management. *JAMA, 285*(19), 2508-2509.

Law, M. (2000). Plant sterol and stanols margarines and health. *British Medical Journal, 320* (7238), 861-864.

Lee, A. F., Bennett, J. A., Manayil, J. C., & Wilson, K. (2014). Heterogeneous catalysis for sustainable biodiesel production via esterification and transesterification. *Chemical Society Reviews, 43*(22), 7887-7916.

Lin, K. M., & Koseoglu, S. S. (2003). Separation of sterols from deodorizer distillate by crystallization. *Journal of Food Lipids, 10*(2), 107-127.

Martins, P. F., Ito, V. M., Batistella, C. B., & Maciel, M. W. (2006). Free fatty acid separation from vegetable oil deodorizer distillate using molecular distillation process. *Separation and Purification Technology*, *48*(1), 78-84.

Masao, N., & Yoshizane, S. (1968) Oryzanol from alkaline cake of rice oil. Japanese Patent 68 12,730 [cited in Chem. Abstr. 69:107773f (1968)].

Mingzhi, L., & Yanyan, L. (1997) Study of oryzanol extracted from the second soapstock of ricebran oil, Chinese Patent 3,300,29 [cited in Chem. Abstr. 130:109308k (1999)].

Moreira, E. A., & Baltanás, M. A. (2004). Recovery of phytosterols from sunflower oil deodorizer distillates. *Journal of the American Oil Chemists' Society, 81*(2), 161-167.

Nagao, T., Watanabe, Y., Nakano, H., & Shimada, Y. (2013). Production and Purification of Functional Lipids Through Enzymatic and Microorganism-Mediated Processes. *Current Organic Chemistry, 17*(8), 776-785.

Narayan, A. V., Barhate, R. S., Indira, T. N., Kaul, P., Raghavarao, K. S. M. S., Appurao, A. G., & Prakash, V. (2005). A simple process for the crystallization of oryzanol from oryzanol enriched fraction, Brazil Patent 200215991.

Narayan, A. V., Barhate, R. S., & Raghavarao, K. S. M. S. (2006). Extraction and purification of oryzanol from rice bran oil and rice bran oil soapstock. *Journal of the American Oil Chemists' Society, 83*(8), 663-670.

Ostlund, R. E. (2002a). Phytosterols in human nutrition. *Annual Review of Nutrition, 22(1)*, 533-549.

Ostlund, R. E., Racette, S. B., Okeke, A., & Stenson, W. F. (2002b). Phytosterols that are naturally present in commercial corn oil significantly reduce cholesterol absorption in humans. *The American Journal of Clinical Nutrition, 75(6)*, 1000-1004.

Panpipat, W., Xu, X., & Guo, Z. (2012). Towards a commercially potential process: Enzymatic recovery of phytosterols from plant oil deodoriser distillates mixture. *Process Biochemistry*, *47*(8), 1256-1262.

Piironen V, & Lampi AM. (2004). Occurrence and levels of phytosterols in foods. In: *Phytosterols as Functional Food Components and Nutraceuticals*. Dutta PC (Ed) Marcel Dekker, Inc. New York, pp 1-32.

Piironen, V., Lindsay, D. G., Miettinen, T. A., Toivo, J., & Lampi, A. M. (2000). Plant sterols: biosynthesis, biological function and their importance to human nutrition. *Journal of the Science of Food and Agriculture, 80(7)*, 939-966.

Pramparo, M., Prizzon, S., & Martinello, M. A. (2005). Study of purification of fatty acids, tocopherols and sterols from deodorization distillate. *Grasas y Aceites*, *56*(3), 228-234.

Pollak, O. J. (1953a). Successive prevention of experimental hypercholesterolemia and cholesterol atherosclerosis in the rabbit. *Circulation, 7*, 696-701.

Pollak, O. J. (1953b). Reduction of blood cholesterol in man. *Circulation, 7*, 702-706.

Rao, K. V. S. A., Rao, B. V. S. K., & Kaimal, T. N. B. (2002) Process for the isolation of oryzanols from rice bran oil soapstock, U.S. Patent 6,410,762.

Ruiz-Méndez, M. V., Márquez-Ruiz, G., & Dobarganes, M. C. (1995). Quantitation of major compounds present in deodorizer distillates from fats and oils. *Grasas y Aceites, 46*(1), 21-25.

Ruiz-Mendez, M. V., Marquez-Ruiz, G., & Dobarganes, M. C. (1996). Comparative performance of steam and nitrogen as stripping gas in physical refining of edible oils. *Journal of the American Oil Chemists' Society*, *73*(12), 1641-1645.

Ruiz-Méndez, M. V., & Dobarganes, M. C. (2007). Combination of chromatographic techniques for the analysis of complex deodoriser distillates from an edible oil refining process. *Food chemistry*, *103*(4), 1502-1507.

Sarup, B., De, M., & Danilo, L. (2014). *U.S. Patent No. 20,140,335,261*. Washington, DC: U.S. Patent and Trademark Office.

Sawadikiat, P., & Hongsprabhas, P. (2014). Phytosterols and γ-oryzanol in rice bran oils and distillates from physical refining process. *International Journal of Food Science & Technology*, *49*(9), 2030-2036.

Seetharamaiah, G. S., & Prabhakar, J. V. (1986). Oryzanol content of Indian rice bran oil and its extraction from soapstock. *Journal of Food Science and Technology*, *23*, 270-273.

Sonntag, N. O. (1990). Fatty Acids, utilization and disposal. *Edible Fats and Oils Processing: Basic Principles and Modern Practices,* 406-412.

Takeshi, Y. (1969) γ-Oryzanol, German Patent 1,301,002 [Cited in Chem. Abstr. 69:128704r (1969)].

Teixeira, A. R. S., Santos, J. L. C., & Crespo, J. G. (2012). Lipase-Catalyzed Consecutive Batch Reaction for Production of Steryl Esters from Vegetable Oil Deodorizer Distillates. *Industrial & Engineering Chemistry Research*, *51*(15), 5443-5455.

Teixeira, A. R. S., Santos, J. L. C., & Crespo, J. G. (2014). Assessment of solvent resistant nanofiltration membranes for valorization of deodorizer distillates. *Journal of Membrane Science*, *470*, 138-147.

Torres, C. F., Reglero, G., & Torrelo, G. (2011). *Extraction and Enzymatic Modification of Functional Lipids from Soybean Oil Deodorizer Distillate.* INTECH Open Access Publisher.

Ubhayasekera, S. J. K. A., & Dutta, P. C. (2009). Sterols and oxidized sterols in feed ingredients obtained from chemical and physical refining processes of fats and oils. *Journal of the American Oil Chemists' Society*, 86, 595-604.

Uddin, M., Sarker, M., Islam, Z., Ferdosh, S., Akanda, M., Haque, J., & Yunus, K. B. (2015). Phytosterols and their extraction from various plant matrices using supercritical carbon dioxide: a review. *Journal of the Science of Food and Agriculture, 95*(7), 1385-1394.

Verleyen, T., Sosinska, U., Ioannidou, S., Verhe, R., Dewettinck, K., Huyghebaert, A., & De Greyt, W. (2002). Influence of the vegetable oil refining process on free and esterified sterols. *Journal of the American Oil Chemists' Society, 79*, 947-953.

Yan, F., Yang, H., Li, J., & Wang, H. (2012). Optimization of phytosterols recovery from soybean oil deodorizer distillate. *Journal of the American Oil Chemists' Society, 89*(7), 1363-1370.

Yang, H., Yan, F., Wu, D., Huo, M., Li, J., Cao, Y., & Jiang, Y. (2010). Recovery of phytosterols from waste residue of soybean oil deodorizer distillate. *Bioresource technology, 101*(5), 1471-1476.

Yin, X., You, Q., Ma, H., Dai, C., Zhang, H., Li, K., & Li, Y. (2015). Biodiesel production from soybean oil deodorizer distillate enhanced by counter-current pulsed ultrasound. *Ultrasonics Sonochemistry, 23*, 53-58.

Winters, R. L. (1990). Deodorizer distillate. In *Proc. World Conf. Edible Fats and Oils Processing, Basic Principles and Modern Practices* Ed. D.R. Erickson, AOCS Champaing, IL, USA, pp. 402-405.

Woerfel, J. B. (1983). Alternatives for processing of soapstock. *Journal of the American Oil Chemists' Society, 60*(2), 310-313.

In: Phytosterols
Editor: Deanna Garner

ISBN: 978-1-63483-477-3


*Chapter 4*

# PHYTOSTEROLS IN OLIVE OIL: OVERVIEW OF ANALYTICAL METHODOLOGY

***Ibrahim M. Abu-Reidah*[1,2,*] *and Amani Taamalli*[3]**

[1] An-Najah National University, Faculty of Science,
Department of Chemistry, Nablus, Palestine
[3]Laboratoire de Biotechnologie de l'Olivier,
Centre de Biotechnologie de BorjCedria, Hammam-Lif, Tunisia

## ABSTRACT

Recent interest in phytosterols stems from their potential protective role, through ingestion of plant foods and oils. This latter approach in turn, results in reducing of serum total cholesterol, and low density lipoprotein (LDL) which may delay or/and prevent occurrence of chronic diseases like CHD and stroke. Many plant foods and vegetable oils contain a variety of phytochemicals with several properties related to individual compounds such as phytosterols which are present in low quantities. For the isolation of phytosterols, high-performance liquid chromatography (HPLC) as well as gas chromatography (GC) coupled with either an ultraviolet/visible (UV/Vis), photodiode array (PDA) and a flame ionization detector (FID) in GC have been widely used. Moreover, mass spectrometry (MS) detector with different analyzers such as ion trap (IT), time-of-flight (Q) TOF, has been hyphenated for identification of phytosterols from diverse vegetable oils. In this sense, the use of HPLC and GC coupled to various detection systems to isolate sterols from olive oil has been reviewed in this chapter. This review is devoted to a short discussion of the occurrence of phytosterols, their role in human health, and focuses on a detailed presentation of the analytical methods, concluding with the advantages of analytical methods employed so far.

**Keywords:** olive oil, phytosterols, HPLC, GC, mass spectrometry

---

* E-mail address: iabureidah@gmail.com; iabureidah@najah.edu.

## INTRODUCTION

Virgin olive oil is obtained from the fruit of the olive tree (*Oleaeuropaea* L.) by using mechanical or other physical means. Olive oil has historically been sought for its gastronomic value and assumed medicinal potentials. Many of the claimed health benefits of olive oil, such as lowering cholesterol concentrations, decreasing the risk of heart disease, promoting insulin levels and blood sugar control are attributed to sterols, which have also been ascribed to have antioxidant, anticancer, antimicrobial and anti-inflammatory properties [1].

The main chemical components of olive oils are triacylglycerols, lower levels of mono-, di-acylglycerols, and complex mixtures of minor compounds (2-3%). The main groups of minor constituents present in olive oil include among others: fatty alcohols, tocopherols, waxes, hydrocarbons, phenolic compounds, volatiles, and phytosterols [2].

According to the IUPAC, sterol molecules consist of four rings with standard carbon numbering marked A, B, C and D, three of which (A, B and C) have six carbon atoms in a nonlinear structure and are attached to one carbon atom ring (D) (Figure 1).

The various phytosterols found in olive oil vary in the number and position of carbon atoms and the double bonds in the side chain and in the ring [3]. Phytosterols are essential structural components of plant membranes. Most phytosterols contain 28 or 29 carbons and one or two carbon-carbon double bonds, characteristically one in the sterol nucleus and sometimes a second in the alkyl side chain.

Figure 2 illustrates the structures of main common olive oil sterols, which show small structural variations. So far, more than 200 sterols and related compounds have been identified, of which about 40 various phytosterols have been found in higher plants; β-sitosterol, campesterol and stigmasterol often predominate [3].

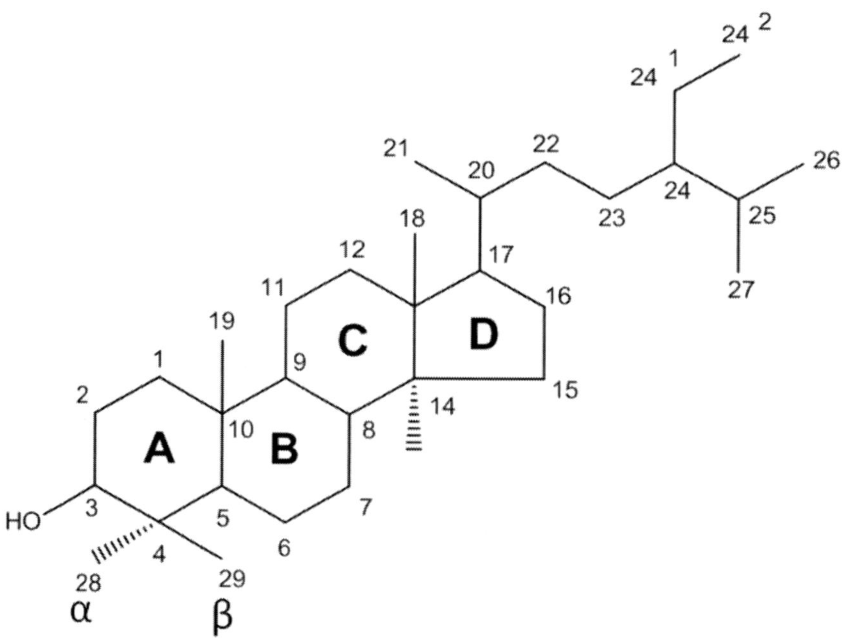


Figure 1. Basic structure of a sterol with standard numbering according to IUPAC system.

Free phytosterols from oils are soluble in alcohols; but insoluble in water. The richest naturally occurring sources of phytosterols are vegetable oils (i.e., Olive oil) and products made thereof. They can be present in the free form and as esters of fatty acid/cinnamic acid or as glycosides [4].

Plant sterols or phytosterols are nutritionally important lipids that need to be routinely determined in foodstuff, especially in vegetable oils. In olive oil, content and composition of sterols can vary as a result of the agronomic, climatic conditions, fruit quality, oil extraction, refining procedures and storage conditions.

Phytosterols make up the greatest proportion of the unsaponifiable fraction olive oils and present a characteristic profile, which make it an important tool for assessing the authenticity of the oil, which is an important issue from the health and commercial perspective.

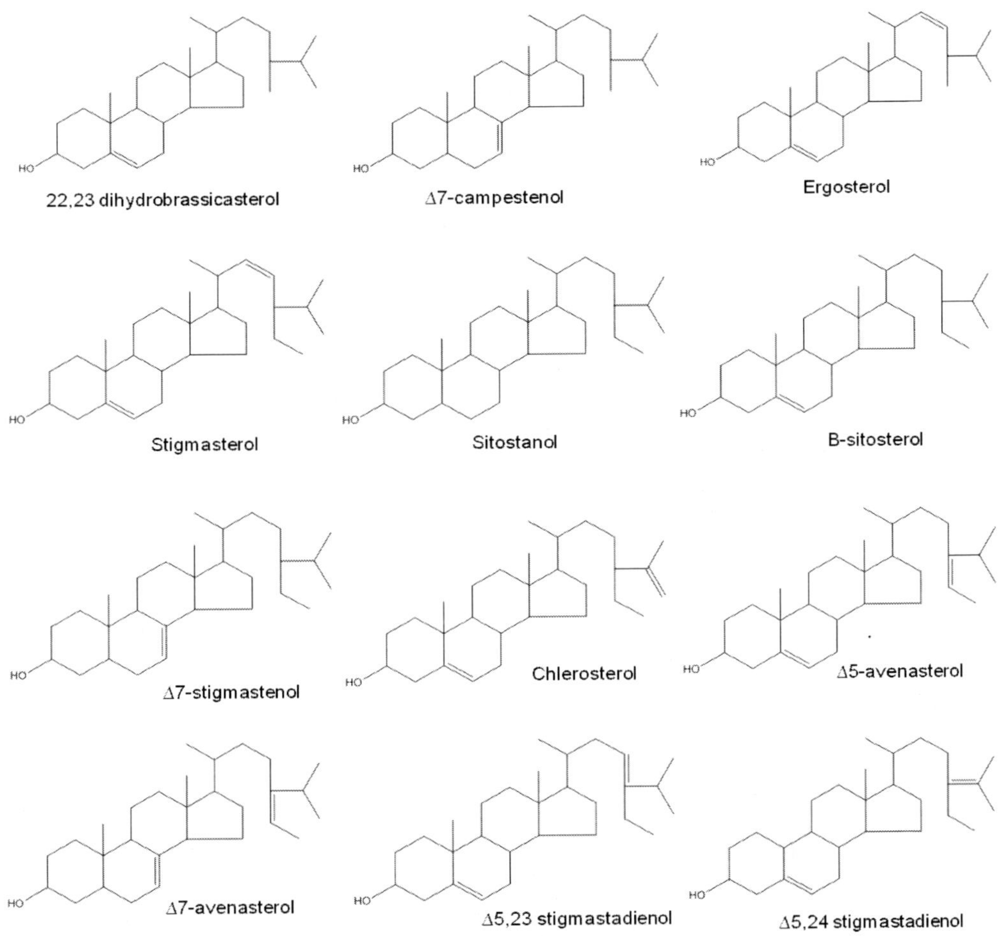


Figure 2. Chemical structures of common sterols present in olive oil.

The quali-quantitative differences in phytosterol composition among several vegetable oils make them a suitable tool to verify botanical origin oils. Moreover, it canbe used to detect adulteration of virgin olive oil with other lower-pricedoils and is one of the criteria internationally used to assess the quality of olive oil [5, 6]. For instance, cholesterol occurs only in very small amounts in plant oils. Besides, olive oil mixed with soybean, grapeseed, or sunflower oil can be detected on the basis of the concentrations of campesterol and stigmasterol. Otherwise, brassicasterol is present in mustard and rapeseed oils but absent or found only in trace amounts in olive oils, and thus is a good identifier of adulteration of olive oil [1, 7]. Nevertheless, sterols have also been used to gauge maturation, quality, type, and geographic region of cultivar used to produce olive oil. Stigmasterol and Δ-5-avenasterol have been shown to increase as the fruit matures, and thus providing a potential measure for determining optimal harvest time [1].

Phytosterols have been reported to present antibacterial, antioxidant, anti-inflammatory and antiulcerative activities and may offer protection against cancers, such as breast, colon and prostate [8]. Also, they actually assist to reduce total plasma cholesterol and LDL cholesterol, and thus, these compounds are considered as functional food ingredients [9]. Additionally, phytosterols can influence the oil stability at high temperatures by acting as polymerization reaction inhibitors [10].

In recent times, it was concluded that the use of phytosterols forlowering blood cholesterol levels in humans was effective and safe. Likewise, the health benefits and safety of phytosterols have already been confirmed by several scholarly reviews [11].

During the last 15 years, there has been a special growth of awareness about phytosterols. Most of this interest has focused on the cholesterol-lowering properties, both dietary and endogenous phytosterols, which bring about a decrease in serum total and low-density lipoprotein cholesterol (LDL-C) [12].

For many years, the occurrence and the dietary effects of these minor sterols were largely ignored and poorly understood. Sterol chemists and biochemists focused their efforts on cholesterol, because elevated serum cholesterol levels were shown to be a prominent risk factor for cardiovascular disease [7].

Olive oil contains common sterols mainly in free and esterified form. Total sterol content (TSC) of virgin olive oils ranges from 1 to 2 g/kg oil. It comprises the largely predominant sterols: sitosterol (between 70 and 95% of total sterols), followed by avenasterol (13%) and campesterol (3%) and 1% of other sterols [4] Other sterols present in smaller quantities or in trace amounts are: stigmasterol, chlerosterol, cholesterol, ergosterol, sitostanol, campestanol, brassicasterol, Δ7-avenasterol, Δ7-cholestenol, Δ7-campestenol, Δ5,23-stigmastadienol, Δ5,24-stigmastadienol, Δ7-stigmastenol, Δ7,22-ergostadienol, Δ7,24-ergostadienol, 22,23-dihydrobrassicasterol, and 24-methylene-cholesterol.

Indeed, lampante olive oils contain higher quantities of total sterols. TSC of solvent extracted olive oils reaches up to three folds higher than that of virgin olive oils. Otherwise, refined olive oils contain total sterols at lower levels, which cause up to 25% of the sterols losses due to the refining process [13].

The common procedure used for sterol analysis in vegetable oils, and other food matrices consists of, a preliminary extraction of the total lipids with an organic solvents, subsequently, saponification and isolation of the unsaponificable fraction, which is then analyzed using various procedures including colorimetric, capillary gas chromatographic (GC), and high performance liquid chromatographic (HPLC) techniques. Among these analytical techniques, GC and HPLC are more selective reliable, and accurate because interference from other components can be easily resolved.

The conventional method for quantifying sterols involves capillary GC analysis with FID, of the fraction isolated by TLC, as trimethylsilyl derivatives. Gas chromatography (GC) with different detectors and with mass spectrometry (MS), are the most usual techniques for sterol determination. High performance liquid chromatography (HPLC) with different detection systems has also been used to determine and quantify major plant sterols in foods. The analysis of sterols in olive oil by HPLC-MS using atmospheric pressure chemical ionization (APCI) as interface was applied for the characterization of the sterol fraction in the olive oil sample. Solid-phase extraction (SPE) has been used to isolate the sterol component from the unsaponifiable fraction [1].

In this chapter, a brief literature survey revealed that most research have favored GC techniques as the methods of choice for the analysis of sterols and related compounds. In some cases, GC appears to deliver greater selectivity for certain isomers than HPLC. The growing public interest in the hypo-cholesterol properties of phytosterols has provided incentive to review existing chromatographic methods for the analysis of sterols in vegetable oils with emphasis on olive oil as important commodity and functional food.

## Sample Preparation and Preliminary Separation

The methods for minor constituent determination usually require isolation and several procedures of separation, identification and quantitation. The analyses are typically based on the official methods developed for determining total sterols in fats and oils. Official methods for isolation of the total sterols from olive oils involve saponification of the lipids, extraction of the non-saponificable matter with organic solvent and washing of the extract with water, separation by thin layer chromatography (TLC) on silica gel plates and derivatization of the sterols and subsequent chromatographic analysis.

Sample preparation is the first and most important step in the analysis of phytosterols. Since they are a minor component that usually comprises less than 1% of the oil matrix, the aims of sample preparation are: to separate the sterol fraction, and to convert all esterified or conjugated phytosterols into free sterols for their subsequent analysis. Sample preparation type needed strongly depends on the sample matrix (i.e., plant extracts, vegetable oils…) and also on the unique form of the phytosterol in the sample. The sample amount is reliant on the concentration of phytosterols in the sample. For instance, samples encompassing around 50 to 150 μg of a mixture of sterols can definitely be analyzed. For analysis of phytosterol, the use of internal standard (IS) is necessary to obtain accurate quali-quantitative results by HPLC or GC. Since retention times often change from run to run, the main purpose of the IS use is to assist in the identification of phytosterols in unknown samples, based on the relative retention time of the phytosterol compared to the IS.

## Saponification Process

In this process, sterol esters are hydrolyzed to free sterols, and the sterols are concentrated in the unsaponifiables' fraction, while the triacylglycerols are converted to fatty acid salts. After saponification, water and/or saturated sodium chloride are added and the unsaponifiable material is extracted by mixing robustly or vortexing with a non-polar solvent such as heptane, chloroform, hexane, diethyl ether, or mixtures thereof. When the phases separate, occasionally by using centrifugation which assists phase separation, the fatty acid salts will partition to the polar phase while the unsaponifiable material will keep on in the a polar phase. After saponification, the unsaponifiables may contain other residues of lipids besides sterols including tocopherols, hydrocarbons, carotenoids, triterpenes, and other free fatty acids. In crude oils such as olive oil, it may be required to further purify the sterol fraction from unsaponifiable polar lipids. Silica and SPE cartridges may be used for this purpose.

## Separation of Phytosterols

Before the development of SPE columns, vegetable oils such as olive, have relatively high content of dimethylsterols. Some of these compounds may co-elute on GC columns with 4-desmethylsterols, requiring the separation into its fractions. Column chromatography and/or TLC may be used to isolate the phytosterol classes. TLC is quicker and easier to run, but low yields can often be obtained by using it. A method developed for separation of the phytosterol classes using silica SPE columns and successive elution with hexane: diethyl ether solvent mixtures of increasing polarity, which resulted in a much higher recovery in comparison with traditional TLC [14].

## Derivatization

Phytosterols are usually analyzed as either sterol acetates or as trimethylsilyl ethers, which in turn improves their response factors, volatility, and peak shape. In fact, free sterols direct injection can result in broader peaks and a lower detector response. The use of acetic anhydride was previously common; nevertheless, silylation is the common derivatization method used these days. The silylation reagent usually used is *N,O*-bis(trimethylsilyl) trifluoroacetamide, often with added 1% trimethylchlorosilane, which acts as a catalyst to improve the reactivity of hampered hydroxyl groups. *N*-methyl-trimethylsilyl-trifluoroacetamide is also utilized, along with other mixtures of hexamethyldisilazane with TMCS (Tri-Sil HTP™ or Hydrox-Sil™). Tri-Sil mixtures are commercially available with differing compositions.

## Separation Techniques of Phytosterols

Due to the metabolic diversity of sterol molecules which often encountered in complex mixtures, demanding the use of chromatographic methods like gas chromatography (GC) and liquid chromatography (LC) for their analyses.

The determination of the composition and content of unsaponifiable material in oils became a very important task during the last decade. Knowledge about the content of sterols, the most interesting part of unsaponifiables, is necessary in order to find new sources of valuable biomolecules, for various food databases realizing people's eating habits, and also for detecting adulteration of virgin olive oils, among others. As the positive effect of sterols against cardiovascular diseases is known, they are widely used as active compounds in various functional foods and medicaments for the treatment and prevention of heart-related diseases. On the other hand, the sterol pattern is a valuable indicator of the quality and origin of butter and similar products. Nowadays the methods of analysis should correspond to many strict requirements like high throughput, reliability, simplicity. One of the most demanding parts of analysis is sample preparation, particularly with matrices like those of vegetable oils, plant matrices, and food products. There are many different procedures available for sample preparation for the determination of sterols in vegetable oils. The most common method for analysis is gas chromatography (GC), which enables precise determination of all types of sterols [15].

## Gas Chromatography (GC)

Actually, chromatographic methods are mostly used for the determination of sterols in olive oil, being GC the most used technique [16, 17, 18, 19].

The Official Method to determine total sterols in fats and oils [20, 21] involves saponification of the lipids with an ethanolic potassium hydroxide solution prior to the analysis by gas chromatography.

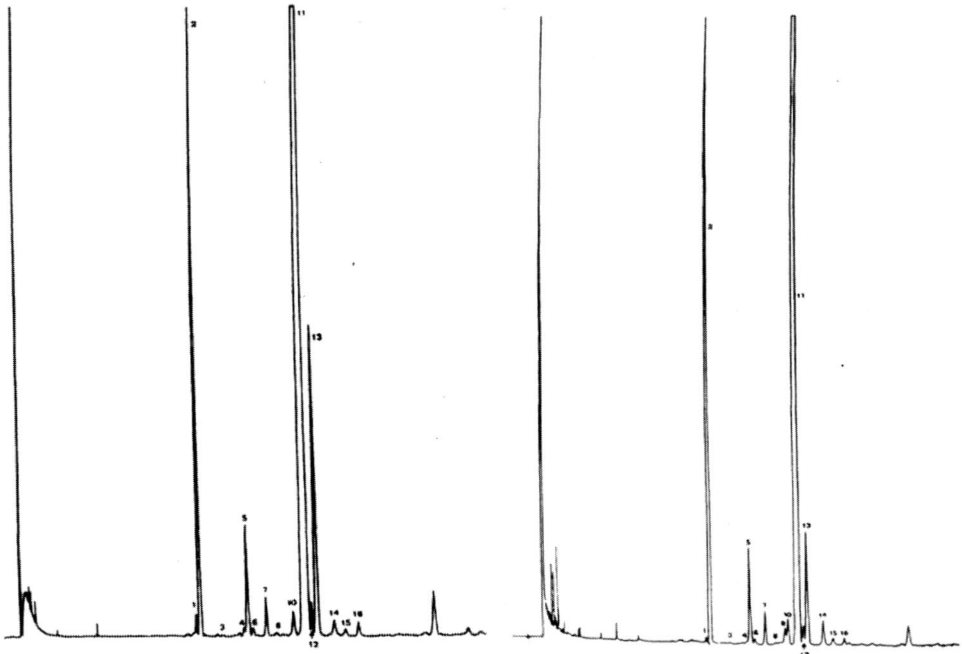

Figure 3. Gas chromatogram of the sterol fraction of crude olive oil (A) and refined olive oil (B) Reproduced from INTERNATIONAL OLIVE OIL COUNCIL, COI/ T.20/ Doc. no.10/ Rev. 1 2001. 1: Cholesterol, 2: Cholestanol, 3: Brassicasterol, 4: 24-methylene-cholesterol, 5: Campesterol, 6: Campestanol, 7: Stigmasterol, 8: Δ-7- campesterol, 9: Δ-5,23-stigmastadienol, 10: Clerosterol, 11: β-sitosterol, 12: Sitostanol, 13: Δ-5-avenasterol, 14: Δ-5-24-stigmastadienol, 15: Δ-7-stigmatenol, 16: Δ-7-avenasterol.

The unsaponifiable matter is then isolated using the diethyl ether extraction that allows the total extraction of sterols, and the extract is washed with water then fractionated by thin layer chromatography (TLC) on basic silica gel plates. Before the injection into the GC for analysis, the derivatization of the sterol fraction is usually performed using a silanizing reagent consisting of a 9:3:1 (v/v/v) mixture of pyridine/hexamethyldisilazane/ trimethylchlorosilane. For the determination of the sterol fraction as trimethylsilyl derivatives, gas chromatography with flame ionization detector (FID) is used as the conventional technique where helium or hydrogen is used as carrier gas. Solid phase extraction (SPE) has been used as alternative of TLC for advantages such as shorter time and lower solvent volume consumption [22]. Figure 3 (A, B) shows chromatograms of sterol fractions fromcrude and refined olive oils obtained by gas chromatography.

For the chromatographic separation of sterols, in the past, packed-glass GC columns were used for phytosterol analysis, but this technology has recently been substituted with fused silica capillary columns due to improvements in resolution and sensitivity. In general, most labs utilize polysiloxane phases of low to mid-polarity.

The most common column phase used for general phytosterol analysis is composed of 95% dimethyl-, 5% diphenyl-polysiloxane [9]. Fused silica capillary column is the most used in olive oil.

It is considered as the column of choice by virtue of its durability and flexibility [24]. Columns most used are 30m in length, 0.25 mm inner diameter, with 0.25µm film thickness. Although columns with 8 or 15m length have been also used (Table 1). Helium was the carrier gas frequently used.

Two-dimensional comprehensive gas chromatography (GC×GC) has been also used to investigate the minor fraction comprising sterol compounds in olive oil [24].

As far as the detectors coupled to the GC for the detection of sterol compounds, FID detector was the most frequently used [17, 21, 25, 26, 20, 27]. Some researches aimed to improve the official method adopted by the IOC and recently, some authors reported an improved and reliable method that permits to analyze thesterol composition more quickly and efficiently than the IOC method with the advantageous use of less solvent, laboratory equipment and manipulation [1].

In addition to FID, mass spectrometry (MS) with electron impact (EI) [21, 22, 27, 28] was also among the most used detectors.

However, few are the studies that reported the use of NMR spectroscopy for the determination of sterols in olive oil [29].

The sterol fraction analyzed in olive oil by GC coupled to several detectors shows the identification of cholesterol, 24-methylencholesterol, campesterol, campestanol, stigmasterol, stigmastanol,7-campesterol, clerosterol, β-sitosterol, sitostanol, avenasterol, 7,9(11)-stigmastadienol, 5,24-stigmastadienol, 7-stigmastenol and 7-avenasterol. Sterol quantification is carried out by adding an internal standard, α-cholestanol [23, 24], which presents a relative retention time to β-sitosterol similar to that of cholesterol. This method can present a drawback. Since when the column ages and loses resolution, these two sterols might coincide in retention time [30].

Among the identified sterols, β-sitosterol, Δ5-avenasterol, and campesterol are the major component in olive oil.

## Table 1. Analysis of sterol fraction by gas chromatography coupled to different detectors

| Samples | Derivatization conditions | Analytical technique conditions | Sterol composition | References |
|---|---|---|---|---|
| 50 virgin olive oils, 50 refined olive oils, and 30 solvent-extracted olive oils, from south of Spain | 250 µL of pyridine-hexamethyldisilazane-trimethyl- chlorosilane (9:3:1, v/v/v) mixture and letting it stand for 15 min. | GC-FID<br>Column: fused silica SPB-5 (30 m x 0.25 mm I.D., 0.25 µm of film thickness)<br>carrier gas: helium<br>Flow rate: 25 mL/min<br>injector temperature, 295°C, detector temperature, 300°C initial oven temperature, 265°C initial time, 35 min; ramp, 20°C final temperature; 300°C final time, 5 min | Identified: cholesterol, campesterol, stigmasterol, Δ5,23-stigmastadienol clerosterol β-sitosterolsitostanol, Δ 5-avenasterol, Δ5, 24-stigmastadienol and Δ7-avenasterol | [30] |
| 11 olive oils from different Mediterranean countries | 200 ml of a pyridine-hexamethyldisilazane–trimethyl-chlorosilane (9:3:1, v/v/v) mixture | GC-FID<br>Column: fused-silica SPB-5 (30 m x 0.25 mm I.D.,0.25 µm film thickness) coated with 5% diphenyl–95% dimethyl-poly-siloxane<br>Oven, injector and detector temperatures: 270, 290 and 310°C, respectively<br>carrier gas: helium<br>Flow rate: 1.6mL/min<br>Split system: 1:50 ratio esterified sterols after column chromatography, 1:10 after SPE fractionation<br>GC-EI-IT/MS<br>Column: fused-silica (30 m x 0.25 mm I.D., film thickness 0.25 µm) coated with poly-(5% diphenyl–95% dimethyl-siloxane)<br>Temperature: 250 to 320°C at a gradient of 3°C/min<br>Injector, transfer line and manifold temperatures: 325, 325 and 220°C, respectively<br>Electron energy : 70eV | Identified: cholesterol, campesterol, stigmasterol,<br>β-sitosterol, β-sitostanol,<br>Δ5-avenasterol Δ7-stigasmtenol | [21] |
| Seven Spanish varieties Carrasquena, Cacerena, Cornezuelo, Corniche, Morisca, Picual, and Verdial | 9:3:1 (v/v/v) mixture of pyridine/hexamethyldisilazane/trimethylchlorosilane | GC-FID<br>Column: TRB-5 (30 m x 0.32 mm I.D., 0.25 µm film thickness)<br>Temperatures: injector 300°C, isothermal column 260°C and detector 325°C<br>Carrier gas: helium<br>Flow rate: 1.1 ml/min | Identified: cholesterol, 24-methylencholesterol, campesterol, campestanol, stigmasterol, 7-campesterol, clerosterol, sitosterol, Sitostanol,<br>5-avenasterol, 5,24-stigmastadienol, 7-stigmastenol and 7-avenasterol<br>Total sterols: Cacerena (1109 mg/kg), Corniche (2030 mg/ kg) | [29] |

**Table 1. (Continued)**

| Samples | Derivatization conditions | Analytical technique conditions | Sterol composition | References |
|---|---|---|---|---|
| | | | β-sitosterol major compound (79.4-86.3 mg/kg) followed by Δ5-avenasterol (4.22-12.33 mg/kg)<br>Campesterol is present at higher levels in the varieties Corniche, Picual, and Verdial (4.01, 3.42 and 3.18; respectively) | |
| Monovarietal olive oils (Cobrançosa, Madural and Verdeal) from Portugal | 1-methylimidazole and N-methyl-N-(trimethylsilyl)-hepta-fluorobutyramide (MSHFBA) | GC-FID<br>Column: DB-5MS<br>(30m × 0.25mm I.D., 0.25 μm)<br>Temperatures of the injector and detector: 320°C<br>Oven temperature: 250°C, increase (2°C/min) to 300°C (12 min)<br>Carrier gas: helium<br>Flow rate: 1.1 ml/min<br>split ratio 1:50 | Identified: cholesterol, campesterol, stigmasterol, β-sitosterol, β-sitostanol, clerosterol, Δ5-avenasterol, Δ7-stigmastenol<br>β-sitosterol: 86.16% (Madural) - 8.34% (Verdeal), Δ5- avenasterol: 6.37% (Verdeal)-8.47% (Madural), campesterol: 2.5% (Madural)- 3.27% (Cobrançosa) | [28] |
| Nine commercial PDO olive oils and 6 monovarietal VOOs from Portugal | 125μL of *N,O*-bis-trimethylsilyltrifluoroacetamid BSTFA (with 1% TMCS) | GC-FID<br>Column: DB-5MS (30m×0.25mm I.D., 0.25μm film thickness) | Identified:<br>Major component in esterified and free fractions: β-sitosterol, followed by campesterol and stigmasterol. | [22] |
| Samples | Derivatization conditions | Analytical technique conditions | Sterol composition | References |
| | and 125μL of pyridine the mixture vortexed for 10 s and heated at 70°C for 20 min | Column temperature: Increase of 2°C/min from 250°C to 300°C for 12 min<br>GC-EI-MS<br>Column: fused silica DB-5MS<br>(30m×0.25mm I.D., 0.25μm film thickness)<br>Colum temperature: 280°C<br>Injector temperature: 270°C<br>Carrier gas: helium<br>Flow rate: 1mL/min<br>Split ratio: 15:1<br>Electron energy : 70eV | In free fractions β-sitosterol: 583.4 to 914.9 mg/kg.<br>Esterified β-sitosterol : 49.8 to 187.1 mg/kg<br>Campesterol : 2.0 to 12.9 mg/kg in free fraction, and 12.6 to 29.7 mg/kg in esterified fraction<br>Stigmasterol <10 mg/kg in free fraction | |

| Samples | Derivatization conditions | Analytical technique conditions | Sterol composition | References |
|---|---|---|---|---|
| 21 varieties cultivated in Turkey | | GC-FID<br>Column: DB-5 (15m x 0.25mm O.D., x 0.1mm I.D., 0.5 mm stationary phase thickness) coated with5%phenyl-/95% methylpolysiloxane<br>Carrier gas: hydrogen | Identified: campesterol, stigmasterol, sitosterol and avenasterol<br>Total sterols: 1.29-2.38 g/kg<br>Sistosterol: 1.03 g/kg (Arbequina)-2.01g/kg (Gökçeada), stigmasterol 0.07 g/kg (Negral)- 0.44 g/kg (Leccino) | [4] |
| Six monovarietal extra virgin olive oils (Regregui, Rekhami, Jarboui, AinJarboua, Neb Jmel and Chétoui) from Northern Tunisia | 9:3:1 (v/v/v) mixture of pyridine/hexamethyldisilazane/trimethylchlorosilane | GC-FID<br>Column: HP-5MS capillary (30 m × 0.25 mm I.D., 0.25 μm film thickness)<br>Temperatures: injector 300°C, isothermal column 260°C and detector 325°C<br>Carrier gas: helium<br>Flow rate: 1.1 ml/min | Identified: cholesterol, 24-M-cholesterol, campesterol β-sitosterol, sitostanol, avenasterol, campestanol, stigmasterol, clerosterol, Δ5,24-stigmastadienol, Δ7-stigmastenol and Δ7-avenastenol<br>Total sterols: 2291.95 mg/kg (Regregui-1287.90 mg/kg (Chétoui)<br>Major sterosl: β-sitosterol, Δ5-avenasterol, and campesterol<br>Regregui cultivar: β-sitosterol (90.24%) with a low level of Δ5-avenasterol (3.78%) | [19] |
| Non-filtered, extra virgin olive oils from Italy | 200 μL of a pyridine:hexam-ethyldisilazane:trimethylchlorosilane mixture (5:2:1, v/v/v) at 40°C for 20 min | GC-FID<br>Column: fused-silica SE52 (30 m x 0.25 mm I.D.,· 0.10 μm film thickness) coated with 5% diphenyl–95% dimethylpolysiloxane<br>Oven temperature: from 240 to 325°C (1.5°C/min)<br>325°C for 10 min<br>Injector and detector temperatures: 325°C<br>Carrier gas: helium<br>Flow rate: 2.7 ml/min<br>Split ratio: 1:30<br>GC-EI-MS<br>Column: ZB-5 fused silica (30 m x 0.25 mm I.D., 0.25 μm film thickness) coated with 5% diphenyl–95% dimethylpolysiloxane<br>Temperature: from 240 to 300°C (3°C/min).<br>The injector, transfer line and manifold temperatures were set at 300, 300 and 200°C, respectively<br>Carrier gas: helium<br>Flow rate: 2.7 ml/min<br>Split ratio: 1:30<br>Electron energy : 70eV | Identified: cholesterol, 24-methylencholesterol, campesterol, campestanol, stigmasterol, stigmastanol, 7-campesterol, clerosterol, β-sitosterol, sitostanol, avenasterol, 7,9(11)-stigmastadienol, 5,24-stigmastadienol, 7-stigmastenol and 7-avenasterol<br>β-sitosterol: 79.77-81.66 %, avenasterol (8.49-9.17%), campesterol (2.88-2.92%) | [31] |

**Table 1. (Continued)**

| Samples | Derivatization conditions | Analytical technique conditions | Sterol composition | References |
|---|---|---|---|---|
| Tunisian cultivar (Chétoui) from different locations | 9:3:1 (v/v/v) mixture of pyridine/hexamethyldisilazane/trimethylchlorosilane | GC-FID<br>Column: HP-5MS capillary column (30 m x 0.25 mm x 0.25 µm)<br>Temperatures: injector 300°C, isothermal column 260°C, and detector 325°C<br>Carrier gas: helium<br>Flow rate: 1.1 ml/min | Identified: cholesterol, 24-methylenecholesterol, campesterol, campestanol, stigmasterol, clerosterol, β-sitosterol, sitostanol, Δ5-avenasterol, Δ5, 24-stigmastadienol, Δ7-stigmastenol and Δ7-avenasterol<br>β-sitosterol: 76.8% (Sers)- 84.4 % (Testour), Δ5-avenasterol: 6.31% (Testour) -16.02 (Sers), campesterol: 2.45% (Bouarada)-2.84 (Testour) | [20] |
| 24 extra virgin olive oils from northern Greece | Phosphorous reagent for NM spectroscopy<br>Hot saponification and silylation for GC analysis | NMR spectroscopy<br>$^{31}$P NMR (Free sterols) and $^{1}$H NMR (Total sterols: free and esterified)<br>High-resolution $^{1}$H NMR spectra acquisition parameters: time domain 32 K; 90o pulse width 9.3 µs; spectral width 10 ppm; relaxation delay 1 s, acquisition time 3.2 s<br>GC-FID<br>Column: DB-5MS fused-silica (30m x 0.25 mm, I.D. 0.25 µm film thickness)<br>Oven temperature: 265°C for 45 min. The injector and detector temperatures were 260 and 295°C, respectively<br>Carrier gas: helium<br>Flow rate: 1.1 ml/min<br>Split ratio: 1:50 | Determined fractions: esterified, free and total sterols<br>Total sterols: 823- 2026 mg/g (NMR), 929-2012 mg/kg (GC)<br>Esterified sterols: 90- 590 mg/g (NMR), 138-501 mg/kg (GC)<br>Free sterols: 720- 1430 mg/g (NMR), 754-1406 mg/kg (GC) | [32] |
| Buza, Crna and Rosinjola varieties from Croatia | pyridine-hexamethyldisilizane-tri-methylchlorosilane (9:3:1, v/v/v) left for 15 min | GC-FID<br>Column: VF-5 ms (30m x 0.25mm I.D. x 0.25 µm film thickness) 5% phenyl:95% dimethylpolysiloxane)<br>Injector, oven and detector temperatures were 280, 260 and 290°C, respectively<br>Carrier gas: helium<br>Flow rate: 1.27 ml/min<br>Split ratio: 1:50 | Identified: cholesterol,<br>24-methylene- cholesterol, campesterol, campestanol, stigmasterol, D7-campes-terol, D5,23-stigmastadienol, clerosterol, β-sitosterol, sitostanol,<br>D5-avenasterol, D5,24-stigmastadienol,<br>D7-stigmastenol,<br>D7-avenasterol) | [23] |

| Samples | Derivatization conditions | Analytical technique conditions | Sterol composition | References |
|---|---|---|---|---|
| Two extra virgin olive oil originated from Italy and Argentina and a hazelnut oil sample fortified with alkyl ester fatty acid | 150 μL of BSTFA and 150 μL of pyridine and left to react for 30 min at room temperature | GC-FID<br>Column: Rxi-5 Sil MS (8 m×0.32 mm I.D.×0.25 μm $d_f$, connected after a deactivated retention gap (60 cmx 0.53 mm I.D.)<br>Oven temperature: 80°C (hold 1min) to 140°C at 20°C $min^{-1}$ then to 350°C (hold 10min) at 5°C $min^{-1}$<br>Carrier gas: helium<br>GC×GC-FID/ EI-MS<br>Columns: Set I Apolar×Mid-polar: A Rxi-5Sil MS column 8 m× 0.25 mm I.D., 0.25 μm $d_f$ (located in GC oven 1) connected to an uncoated capillary segment<br>(1.5 m× 0.25 mm I.D.) and then to a Rxi-17Sil MS segment (1.5 m×0.15 mm i.d., 0.15 μm $d_f$) located in the second oven.<br>Set II Mid-polar×Apolar: A Rxi-17Sil MS column 8 m× 0.25 mm I.D., 0.25 μm $d_f$ (located in GC oven 1) connected to an uncoated capillary segment (1.5m×0.25mm I.D.) and then to a Rxi-5SilMS segment (1.5m×0.15mmi.d., 0.15 μm $d_f$) located in the second ovenOven temperature program: 50°C (hold 1min) to 140°C at 40°C $min^{-1}$, then to 350°C (hold 5 min) at 4°C$min^{-1}$ in the first oven, and 50°C (hold 1min) to 165°C at 40°C $min^{-1}$ then to 360°C(hold 8min) at 4°C $min^{-1}$ in the second oven<br>Carrier gas: helium<br>Injection temperature: 300°C | Identified: TMS-campestanol, stigmasterol, β-sitosterol, Δ5-avenasterol and cycloartenol<br>β-sitosterol: 145.19 mg/100g (Italian olive oil) - 169.01 mg/100g (Argentinian olive oil) | [27] |

**Table 2. Analysis of sterol fraction by liquid chromatography coupled to different detectors**

| Samples | Sample preparation | Analytical technique | Sterol composition | References |
|---|---|---|---|---|
| Extra-virgin olive oil samples | EVOO was added to KOH ethanolic solution, heated at 80°C until saponification. Distilled water was added and cooled. After adding diethyl ether, the solution was vortex-mixed. Next, centrifuged, then separately collected and treated as follows: The aqueous phase was treated with diethyl ether.<br>Diethyl ether solution was dried in vacuo. Non-saponificable material was dissolved in chloroform and passed through a Sep-Pak Vac C18 cartridge. | Nano-LC-UV–vis<br><br>Column:<br>RP-C18 sub-2 μm,<br><br>Mobile phase: MeOH<br>Flow rate: 330 nL/min.<br><br>MS conditions:<br>LCQ quadrupole ion-trap mass spectrometer with (nano-ESI) source fused silica capillary (10.5 cm × 25 m I.D. × 375 m O.D.)<br><br>Positive ion mode.<br>The ion-spray voltage: 2.0 kV capillary temperature : 200°C<br>The mobile phase was improved with ammonium acetate salt. Time of analysis: 20 min. | Brassicasterol,<br>Stigmasterol,<br>Campesterol, Cholesterol<br>β-sitosterol | [31] |
| Olive oil (Richfood) | Oil samples were mixed with isopropanol and then saponified in ethanolic KOH at 80 C for 60 min. Deionized water and n-hexane were added, and the mixture was vortex mixed. After centrifugation, the hexane layer was collected, and then evaporated to dryness. The residue was reconstituted in isopropanol for analysis using LC–MS–MS. | LC–MS–MS<br><br>Column: Zorbax XDB-C18 column (2.1x50 mm, 5 μm).<br><br>MS condition:<br>API triple quadrupole mass spectrometer<br>Positive ion atmospheric pressure chemical ionization (APCI)<br>Mobile phase: Isocratic acetonitrile/methanol (99:1, v/v).<br><br>Flow rate: 600 μL/min,.<br>The ion source temp.: 300◦C. | Brassicasterol<br>Campesterol Cycloartenol<br>Lupeol<br>β-Sitosterol Stigmasterol | [32] |

| Samples | Sample preparation | Analytical technique | Sterol composition | References |
|---|---|---|---|---|
| Virgin olive oil Refined olive oil Olive-pomace oil Crude olive-pomace oil samples | Olive oil samples were saponified by refluxing with ethanolic KOH solution. After cooling, water was added. Afterward separation of aqueous phase, it was washed with diethyl ether. Then, the diethyl fractions were washed with water and dried, filtered, evaporated to dryness at reduced pressure, and the unsaponificable was dissolved in chloroform. The extract was subjected to TLC on silica gel plates. | LC-MS<br>Mobile phase: acetonitrile/acidified water<br>Flow rate: 0.5 mL/min. Column : 5-μm Atlantis d C18 (150x 2.1-mm i.d.)<br>Temp,: 30° C Injection volume: 10 μL.<br><br>MS conditions:<br>Positive ion mode.<br>The temperature source was 120 C.<br>The flow cone gas was 350 L/h. | Cholesterol Stigmasterol<br>β-Sitosterol<br>Sitostanol<br>Fucosterol<br>Erythrodiol–uvaol | [33] |
| Six Tunisian monovarietal virgin olive oils | Olive oil samples were saponified with ethanolic KOH solution. Distilled water was added. After phase separation aqueous phase was washed with diethyl ether. Then diethyl ether fractions were washed with water and with ethanolic KOH solution, then dried with anhydrous sodium sulfate. Filtered and evaporated to dryness using a rotary evaporator, and the residue was dissolved in acetonitrile. The next racts werefiltered through (0.45 μm) filter. | HPLC-APCI-IT MS/MS<br>LC-MS system comprised of HPLC equipped with diode array detector module, with a column Phenomenex C18 (5 μm, 3.0x250 mm)coupled with Bruker Esquire 2000 ion trap with an APCI interface. The mobile phase flow was at 0.5 mL/min. acquisition of MS/MS data. The ionization was positive-ion APCI, and mass spectra were acquired up to mass m/z using the following parameters: temperature: 350 C and flow rate of nitrogen was 5 L/min, and the APCI interface temperature was 450°C.<br>The pressure of the nebulizer gas was set to 50 psi, and the voltage of the corona discharge needle was 3500 V. Capillary voltage was set at -3000 V and the end-plate offset at -500 V. | Erythrodiol<br>Uvaol<br>Lupeol<br>Brassicasterol<br>Cholesterol<br>Δ5-avenasterol<br>Campesterol<br>β-amyrin Stigmasterol<br>β-sitosterol | [34] |

**Table 2. (Continued)**

| Samples | Sample preparation | Analytical technique | Sterol composition | References |
|---|---|---|---|---|
| Olive oil and other cooking oils | Derivatization of oil samples performed with d0/d5-pyridine.<br>Oil sample was Saponification by ethanolic potassium hydroxide solution by refluxing. The unsaponifiable fraction was extracted with ethyl ether, the extract dried, Oil extract was transferred and added with pyridine in dichloromethane or<br>d5-pyridine in dichloromethane then mixed with $Tf_2O$ in dichloromethane, vortexed, d0-labeled derivative was combined with its d5-labeled, diluted with methanol for ESI-Q-TOF MS analysis. | HPLC -Q-TOF MS<br>Mobile phase : H2O: CH3CN 50:50,<br>Constant flow rate:0.2 ml/min<br>Injection volume: 1 μL.<br><br>MS conditions :<br>Positive (+) ESI mode.<br><br>The gas temp.: 350°C Drying gas: 8 L/min<br>Nebulizer : 35 psi Capillary voltage : 4 KV<br>Fragmentor : 120 V<br>Skimmer :45 V<br>Spectral mass range : 350–800 m/z.<br>Collision energy : 30 eV. | CholesterolDihydrocholesterol Stigmasterol<br>β-sitosterol<br>Lanosterol | [35] |
| 25 samples of Virgin olive, 5 samples of pomace olive and other vegetable oils | Saponification by ethanolic potassium hydroxide solution by refluxing. The unsaponifiable fraction was extracted with ethyl ether, the extract dried, evaporated to dryness in a rotary evaporator. | HPLC-UV<br>Column: Lichrospher 100 CN column<br>(250 mm × 4.0 mm i.d., 5 μm), Mobile phase: hexane:TBME, 80:20 (v/v).<br>Flow rate: 0.8 mL/min Detection wavelength: 208 nm. | - | [36] |
| 3 Extra virgin olive oils and otheroils | Oil was mixed with KOH in ethanol. The mixture was heated under reflux. After cooling water and diethyl ether were added. After separation in a separation funnel, the aqueous phase was washed with diethyl ether. Then extract was evaporated to dryness. The remaining unsaponificable material was dissolved in methanol, filtered. | RPLC–SALDI MS<br>The detectionwavelength : 206 nm.<br>Column :Zorbax (SB-C18 100× 2.1 mm I.D., 3.5 μm)<br>Mobile Phases: methanol–water.<br>Flow rate of 0.7 mL/min. Injection volume was 20 μL.<br>Mass range: m/z 400–600 in the positive ion mode.<br><br>MS conditions : accelerating voltages 18.99 kV and 16.69 kV, suppression of ions with m/z < 400, spectra averaging from 2000 laser shots (20 ×100 shots) per spot and laser power adjusted to 10–60%, typically 45%. | Brassicasterol Cholesterol Stigmasterol Campesterol B-sitosterol | [37] |

| Samples | Sample preparation | Analytical technique | Sterol composition | References |
|---|---|---|---|---|
| virgin olive, olive-pomace oils, other vegetable oils | After Saponification with KOH in ethanol solution,<br>the unsaponifiable fraction of diethyl ether<br>is collected, washed with water and dried with anhydrous<br>sodium sulfate. Then filtered and evaporated to dryness using a rotary evaporator. The residue is dissolved in methanol. Then filtered<br>through a membrane filter (0.45 mm) before analysis. | HPLC-APCI-MS<br>Column: Waters Atlantis 5mm dC18 150x2.1 mm<br>Gradient: acetonitrile/acidified water<br>Flow rate: 0.5 mL/min<br>Column temperature: 30˚C.<br>MS conditions were:<br>an ebulizer temp.: 450˚C, Source temp. 120 ˚C.<br>Gas flow rate of 50 L/h; and a desolvation gas flow rate of 350 L/h.<br>Injection volume:10 mL | Erythrodiol<br>Uvaol<br>Stigmastadienol<br>24-Methylene Cholesterol<br>Brassicasterol<br>Δ7-Avenasterol Cholesterol<br>Δ5-Avenasterol<br>Δ7-Campesterol Clerosterol<br>Campesterol Campestanol<br>Stigmasterol<br>Δ7-Stigmastenol<br>Δ5,24-Stigmastadienol<br>β-Sitosterol Sitostanol | [38] |
| Five samples of VOOs | Hot saponification<br><br>Cold saponification<br><br>Microwave-assisted saponification<br><br>Ultrasound-assisted saponification | LC–TOF/MS<br>Column: C18 Pursuit XRs Ultra (50mmx2.0mm id, 2.8mm particle size)<br><br>Mobile phase:<br>was: A (0.1% formic acid in water) and B (0.1% formic acid in ACN).<br>Flow rate: 0.5 mL/min.<br>Injection vol.: 5mL.<br><br>The MS conditions:<br>Gas temperature, 350°C; Drying gas flow at 10 L/min.<br><br>Nebulizer pressure, 35 psi.<br><br>Sheath gas temperature, 380°C.<br>Sheath gas flow:10 L/min;<br>Capillary voltage, 3250V.<br>Positive ionization mode Skimmer: 65 V.<br>The mass range: m/z 100–1100. | Δ5 avenasterol<br>Campesterol | [39] |

## High-Performance Liquid Chromatography(HPLC)

By the introduction of the HPLC technology in latest years, researchers have used HPLC methods for the analysis of sterols (Table 2). HPLC has bettered GC in one area that it operates under slighter column temperature, in addition to undamaging detection conditions. Consequently, HPLC technique has ideally well-matched the analysis of thermally unstable sterols. At times, direct HPLC analysis of sterols can be achieved with little sample purification to prevent undesirable sample losses. Otherwise, samples of complex matrices must go through other chromatographic pretreatment such as Column Chromatography (CC) and/or Thin layer Chromatography (TLC) to get rid of interfering endogenous matters. Hence, employing HPLC procedures are much constrained by the complexity of sample matrices, the type and concentration of sterol components. Since the pioneering HPLC separations of sterols, normal-phase and reversed-phase HPLC techniques have been widely used for the analysis of individual sterols in various sample matrices. In general, HPLC analyses of sterol components have more often been carried out in the reversed phase mode.

## Detection Techniques

The commercially available HPLC detection methods for the detection of sterols include: evaporating light scattering detector (ELSD), photodiode array detection (PAD), Ultraviolet (UV), refractive index detection (RI), and mass spectrometry (MS) systems.

Of these, UV (200-210 nm settings) and RI detectors are the most useful for sterol assays, with the former providing better detection sensitivity. Because the number and location of double bonds in sterols differ widely, HPLC quantification of sterols with UV detection is less than straightforward owing to the variable UV absorption. Therefore, calibration plots of UV detector against standards must be constructed prior to HPLC sample quantitation.

In HPLC-DAD system, analytes emerging from column effluents are monitored with multiple-wavelength UV diode array detector scanning from 200 to 400 nm suitable for the selective analysis of sterols with various conjugated olefinic structures. Although plant sterols in vegetable oils are rarely analyzed by HPLC-fluorescence (FL) detection, the FL detection technique can be valuable in trace analyses of fluorescence-labeled sterols with enhanced detection sensitivity and specificity.

Coupling HPLC with MS or NMR not only will permit simultaneous separations and quantitation but also will provide a valuable means for structural elucidation and confirmation of sterol structures. However, few publications on the practical application of HPLC-MS and HPLC-NMR techniques in sterol analysis have appeared in the literature. In some methods, the sterol fractions obtained from HPLC were characterized by GC-MS, whilst the individual sterols isolated from HPLC were analyzed by MS or NMR.

## Other Chromatographic Techniques

Capillary electrochromatography (CEC) is a chromatographic tool in which the mobile phase is driven over the chromatographic bed by electroosmosis. It was first introduced in

1970th and is a relatively new technique for the analysis of sterols. CEC combines principles used in HPLC and CE. The CEC reversed-phase elution patterns are comparable with HPLC.

Supercritical fluid chromatography (SFC) employs inert supercritical carbon dioxide as mobile phase eluent. The technique is suitable for the analysis of phytosterols in vegetable oils. Analytical precision and sensitivity of the various techniques currently available for sterol assays seems to follow the order GC > HPLC > SFC > CEC. The sensitivity order may differ depending on sterol structures and detectors coupled to chromatographic instruments. While GC, HPLC are individually meritorious for specific applications, GC-FID (or MS) is considered to be the method of choice for practical analyses of plant sterols in foods and vegetable oils.

## CONCLUSION

Phytosterols encompass a major proportion of the non-saponifiable components of olive oil. They are important from a nutritional point of view. Phytosterols analysis can be used to detect olive oil adulteration with other low-priced vegetable oils. In this chapter, different powerful analytical tools, viz. GC and HPLC coupled with PAD, MS and NMR detection systems have been described for the separation and determination of phytosterols in olive oil. These techniques have proved very useful for determining most of unknown compounds in different olive oil samples. The difficulty to obtain standards and the spectral similarities of the phytosterols represents a major weakness. Therefore, MS and NMR have become the preferred techniques for phytosterols characterization. Given the beneficial effect of phytosterols for the health such as antioxidant, hypo-cholesterol, anti-inflammatory and antimicrobial activities, and their use as a natural additive in processed and functional foods, the interest in phytosterols continues to grow. The implementation of better extraction, purification, and identification methodologies will have an impact in the establishment of new tools for the food-product quality control and authentication. The analytical methodologies described herein represent the current methodologies for the analyses of phytosterols. According to the sensitivity and precision of chromatographic methods, the currently available techniques for sterol analysis can be categorized as follows: GC > HPLC > SFC. GC–FID/(MS) can be considered the method of choice for the estimation of phytosterols in olive oil. Otherwise, other procedures using HPLC coupled with various detectors are to be developed to facilitate the analysis of phytosterols in olive oil.

## REFERENCES

[1] Mathison, B., Holstege, D. A rapid method to determine sterol, erythrodiol, and uvaol concentrations in olive oil. *J. Agric. Food Chem.* 2013, *61*, 4506–4513.

[2] Abu-Reidah, I. M., Yasin, M., Urbani, S., Servili, M., Montedoro, G. Study and characterization of Palestinian monovarietal Nabali virgin olive oils from northern West Bank of Palestine. *Food Res. Int.* 2013, *54*, 1959-1964.

[3] Gül, M. K., Şeker, M. Comparative analysis of phytosterol components from rapeseed (Brassica napusL.) and olive (OleaeuropaeaL.) varieties. *European Journal of Lipid Science and Technology* 2006, *108*, 759-765.

[4] http://en.wikipedia.org/wiki/Phytosterol.(accessed April 04,2015).

[5] Amaral, J. S., Mafra, I., Oliveira, M. B. P. P. In *Chapter 63 - Characterization of Three Portuguese Varietal Olive Oils Based on Fatty Acids, Triacylglycerols, Phytosterols and Vitamin E Profiles: Application of Chemometrics,* Watson, V. R. P. R., Ed., Olives and Olive Oil in Health and Disease Prevention, Academic Press: San Diego, 2010, pp 581-589.

[6] Azadmard-Damirchi, S. Review of the use of phytosterols as a detection tool for adulteration of olive oil with hazelnut oil. *Food Additives & Contaminants: Part A* 2010, *27*, 1-10.

[7] BohacenkoI., Kopicova, Z. Detection of Olive Oils athenticityby determination of their sterol content using LC/GC. *Czech J. Food Sci.* 2001, *19*, 97-103.

[8] Awad, A. B., Fink, C. S. Phytosterols as Anticancer Dietary Components: Evidence and Mechanism of Action. *The Journal of Nutrition* 2000, *130*, 2127-2130.

[9] Aldini, R., Micucci, M., Cevenini, M,, et al. Antiinflammatory Effect of Phytosterols in Experimental Murine Colitis Model: Prevention, Induction, Remission Study. *PLoS ONE* 2014,*9*, e108112.

[10] Dobarganes, M. C., Velasco, J. Analysis of lipid hydroperoxides. *Eur. J. Lipid Sci. Tech.* 2002, *104*, 420-428.

[11] Kritchevsky, D., Chen, S. C. Phytosterols—health benefits and potential concerns: a review. *Nutr. Res.* 2005, *25*, 413-428.

[12] Pascal, L., Segal, R. Phytosterols – Biological active compounds in food. *J. Agroaliment. Proc. Technol.* 2006,*12*, 149-158.

[13] Pokorný, J. Olive Oil.Chemistry and Technology.Edited by D. Boskou. VII. AOCS Press, Champaign, Illinois, 1996. *Food/Nahrung* 1997, *41*, 119-120.

[14] Azadmard-Damirchi, S., Nemati, M., Hesari, J., Ansarin, M. and Fathi-Achachlouei, B. Rapid separating and enrichment of 4,4'-dimethylsterols of vegetable oils by solid-phase extraction. *J Am. Oil Chem. Soc.* 2010, *87*, 1155-1159.

[15] Mäeorga, E., Läänistea, P., Jõudua, J., Mäeorgb, U. Some important aspects of sterol analysis of vegetable oils. *Proc. Estonian Acad. Sci. Chem.* 2007, *56*, 59–66.

[16] Haddada, F. M., Manaï, H., Oueslati, I., Daoud, D., Sánchez, J., Osorio, E., Zarrouk, M. Fatty acid, triacylglycerol, and phytosterol composition in six Tunisian olive varieties. *J. Agric. Food Chem.* 2007, *55*, 10941–10946.

[17] Ben Temime, S., Manai, H., Methenni, K., Baccouri, B., Abaza, L., Daoud, D., Casas, J., Bueno, E., Zarrouk, M. Sterolic composition of Chétoui virgin olive oil: Influence of geographical origin. *Food Chem.* 2008, *110*, 368–374.

[18] Cercaci, L., Rodriguez-Estrada, M. T., Lercker, G. Solid-phase extraction-thin-layer chromatography-gas chromatography method for the detection of hazelnut oil in olive oils by determination of esterified sterols. *J. Chromatogr. A* 2003, *985*, 211–220.

[19] Cunha, S. S., Fernandes, J. O., Oliveira, M. B. P. P. Quantification of free and esterified sterols in Portuguese olive oils by solid-phase extraction and gas chromatography-mass spectrometry. *J. Chromatogr. A* 2006, *1128*, 220–227.

[20] International Olive Oil Council, Methods of analysis. Determination of the compostion and content of sterols by capillary-column gas chromatography. *COI/T20/Doc.no.10/ Rev1, Madrid* 2001, 1–16.

[21] Official Journal of the European Union Regulation EEC N° 2568/91 on the Characteristics of olive oil and olive-residue oil and on the relevant methods of analysis. http://eur-lex.europa.eu/legal-content/EN/TXT/?uri=CELEX:32013R0299. (accessed on June 05, 2015).

[22] Abidi, S. L. Chromatographic analysis of plant sterols in foods and vegetable oils.*J. Chromatogr. A* 2001, *935*, 173–201.

[23] Lukić, M., Lukić, I., Krapac, M., Sladonja, B., Piližota, V. Sterols and triterpene diols in olive oil as indicators of variety and degree of ripening. *Food Chem.* 2013, *136*, 251–258.

[24] Purcaro, G., Barp, L., Beccaria, M., Conte, L. S. Fingerprinting of vegetable oil minor components by multidimensional comprehensive gas chromatography with dual detection. *Anal. Bioanal. Chem.* 2014, *407*, 309–319.

[25] Alves, M. R., Cunha, S. C., Amaral, J. S., Pereira, J. a., Oliveira, M. B. Classification of PDO olive oils on the basis of their sterol composition by multivariate analysis. *Anal. Chim. Acta* 2005, *549*, 166–178.

[26] Sanchez casas, J., Osario Bueno, E., García Montaño, A., Martinez Cano, M. Sterol and erythrodiol+uvaol content of virgin olive oils from cultivars of Extremadura (Spain). *Food Chem.* 2004, *87*, 225–230.

[27] Bias, D., Gonz, V. Determination of Sterols by Capillary Column Gas Chromatography. Differentiation Among Different Types of Olive Oil: Virgin, Refined, and Solvent-Extracted. *J. Am. Oil Chem. Soc.* 1996, *12*, 1685–1689.

[28] Cercaci, L., Passalacqua, G., Poerio, A., Rodriguez-Estrada, M. T., Lercker, G. Composition of total sterols (4-desmethyl-sterols) in extravirgin olive oils obtained with different extraction technologies and their influence on the oil oxidative stability. *Food Chem.* 2007, *102*, 66–76.

[29] Hatzakis, E., Dagounakis, G., Agiomyrgianaki, A., Dais, P. A facile NMR method for the quantification of total, free and esterified sterols in virgin olive oil. *Food Chem.* 2010, *122*, 346–352.

[30] Aparicio-Ruiz, R., Harwood, J., *Handbook of Olive Oil: Analysis and Properties*. Springer.

[31] Rocco, A., Fanali, S. Analysis of phytosterols in extra-virgin olive oil by nano-liquid chromatography. *J Chromatogr A* 2009, *1216*, 7173-7178.

[32] Mo, S., Dong, L., Hurst, W. J., van Breemen, R. B. Quantitative Analysis of Phytosterols in Edible Oils Using APCI LiquidChromatography–Tandem Mass Spectrometry. *Lipids* 2013,*48*, 949–956.

[33] Cañabate-Díaz, B., Segura Carretero, A., Fernández-Gutiérrez, A., Belmonte Vega, A., GarridoFrenich, A., Martínez Vidal, J. L., Duran Martos, J. Separation and determination of sterols in olive oil by HPLC-MS. *Food Chem.* 2007, *102*, 593-598.

[34] Zarrouk, W., Carrasco-Pancorbo, A., Segura-Carretero, A., Fernández-Gutiérrez, A., Zarrouk, M. Exploratory Characterization of the Unsaponifiable Fraction of Tunisian Virgin Olive Oils by a Global Approach with HPLC-APCI-IT MS/MS Analysis. *J. Agric. Food Chem.*2010, *58*, 6418–6426.

[35] Wang, H., Wang, H., Zhang, L., Zhang, J., Leng, J., Cai, T., Guo, Y. N-Alkylpyridinium Quaternization for Assisting Electrospray Ionization of Sterols in Oil by Quadrupole-Time of Flight Mass Spectrometry. *J. Mass Spectrom.* 2013, *48,* 1101–1108.

[36] Bagur-González, M. G., Pérez-Castaño, E., Sánchez-Viñas, M., Gázquez-Evangelista, D. Using the liquid-chromatographic-fingerprint of sterols fraction to discriminate virgin olive from other edible oils. *J Chromatogr A* 2015, *1380*, 64-70.

[37] Vrbková, B., Roblová, V., Yeung, E. S., Preisler, J. Determination of sterols using liquid chromatography with off-line surface-assisted laser desorption/ionization mass spectrometry. *J Chromatogr A* 2014, *1358*, 102-109.

[38] Carretero, A. S., Carrasco-Pancorbo, A., Cortacero, S., Gori, A., Cerretani, L., Fernández-Gutiérrez, A. A simplified method for HPLC-MS analysis of sterols in vegetable oil. *Eur J Lipid Sci Tech* 2008, *110*, 1142-1149.

[39] de Medina, V. S., Priego-Capote, F., Luque de Castro, M. D. Comparison of saponification methods for characterization of the nonsaponifiable fraction of virgin olive oil. *European Journal of Lipid Science and Technology* 2013, *115*, 1325-1333.

In: Phytosterols
Editor: Deanna Garner

ISBN: 978-1-63483-477-3


*Chapter 5*

# PHYTOSTEROLS: APPLICABILITY AS CHOLESTEROL METABOLISM BIOMARKERS AND HEALTH BENEFITS

***Isabel Andrade*[1,2], *Lèlita Santos*[3] and *Fernando Ramos*[2,4,*]**

[1]Instituto Politécnico de Coimbra, ESTESC-Coimbra Health School, Ciências Complementares, Rua 5 Outubro, S. Martinho do BispoCoimbra, Portugal
[2]CEF-Center for Pharmaceutical Studies, Health Sciences Campus, Pharmacy Faculty, University of Coimbra, Azinhaga de Santa Comba, Coimbra, Portugal
[3]Department of Internal Medicine, Coimbra University Hospitals, Medicine Faculty, University of Coimbra, Coimbra, Portugal
[4]CNC-Center for Neuroscience and Cell Biology, Pharmacy Faculty, University of Coimbra, Azinhaga de Santa Comba, Coimbra, Portugal

## ABSTRACT

The consumption of phytosterols (plant sterols and/or stanols) can reduce cardiovascular risk, by reducing hypercholesterolemia. Serum cholesterol concentrations can be determined by the traditional lipid profile parameters (total cholesterol, LDL-C, HDL-C, triglyceride), and in addition, the cholesterol metabolism can also be studied by the quantification of the intestinal cholesterol absorption and endogenous cholesterol synthesis markers. The disruption of the cholesterol homeostasis between absorption and synthesis of cholesterol is related to a positive history of several cardiovascular diseases and will be focused in this chapter.

Sitosterol and campesterol are the major sources of phytosterols in diet, and are therefore the main biomarkers used to study the exogenous cholesterol absorption, while desmosterol and lathosterol, the late precursors of the cholesterol biosynthetic pathway, are the most common biomarkers used to quantify cholesterol synthesis, whether in diet/lifestyle interventions or drug therapy monitoring.

There is evidence that several food matrices enriched with phytosterols, or supplements, are effective in lowering serum LDL-C concentrations, when used alone as part of the diet or when used as an adjuvant to ongoing statin therapy. In addition, the market for functional foods and dietary supplements with a health claim (for lipid-lowering) is expanding rapidly worldwide, and consequently an increasing number of

* Corresponding author: framos@ff.uc.pt.

persons will use these products and combine them with their prescribed drugs. Therefore, several studies have investigated the effect on lipids and also on serum cholesterol metabolism markers of a variety of phytosterol-enriched food matrices in hypercholesterolemic patients and in statin-treated individuals.

In addition to the hypocholesterolemic effect of the phytosterol-enriched food products, it is considered that in the particular case of the fermented dairy (low-fat) products, there are other benefits, such as nutritional, physiological and immunological positive overall effects on health, and in particular, in intestinal health.

In sum, this chapter will include a description of phytosterol-enriched food matrices studied/used to lower cholesterol levels and developments and trends in the field will be discussed. Studies on the use of these functional foods in hypercholesterolemic patients, on and off drug therapy, will be referenced. The use of the cholesterol metabolism markers will be highlighted in regard to the study of cardiovascular diseases, and the potential role of phytosterols will be discussed in disease prevention and health promotion.

**Keywords:** cardiovascular diseases, cholesterol, campesterol, sitosterol, desmosterol, lathosterol

## 1. Phytosterols as Cholesterol-Lowering Agents

A positive history of several atherosclerotic cardiovascular diseases (CVD) is related to the disruption of cholesterol metabolism homeostasis among other modifiable (lifestyle, hypertension, diabetes, obesity, and metabolic syndrome) and non-modifiable (sex and age) risk factors. Increased blood cholesterol is an independent risk factor for atherosclerotic CVD. Early lowering of low-density lipoprotein cholesterol (LDL-C) may delay or prevent the onset of atherosclerosis. Therefore, total cholesterol (TC) and LDL-C serum levels have been the primary targets of hypercholesterolemia therapy, with the intent of reducing cardiovascular risk (Reiner et al., 2011).

To avoid cholesterol accumulation in serum, tissues or biliary tract resulting in pathological conditions, there's a tight regulation of the main components of the whole-body cholesterol metabolism, including exogenous intestinal absorption, endogenous synthesis, cholesterol trafficking to and from tissues and excretion from the body.

Both the liver and intestine are central to human cholesterol homeostasis.

The intraluminal cholesterol absorbed in the proximal small intestine (50% is transported to the liver) is derived from diet (20%), from bile (60%) and also from enterocytes (20%). The hepatic *de novo* cholesterol synthesis is about 800 mg/day and 300 mg of the cholesterol pool is synthesized to bile acids (Turner et al., 2012).

Cholesterol absorption is achieved primarily through its incorporation into (mixed) micelles that allow hydrophobic molecules like cholesterol to contact with the apical membrane of the enterocytes. Niemann-Pick C1-Like 1 (NPC1L1) is the transmembrane protein localized at the apical membrane of enterocytes that functions as a sterol transporter, mediating the intestinal cholesterol absorption and counter-balancing hepatobiliary cholesterol excretion. Cholesterol that enters the enterocytes is then packaged into chylomicrons for secretion into lymph and blood via the ATP-binding cassette sub-family G types 5 and 8 (ABCG5/8), a heterodimer system of transport located at the enterocyte's basolateral membrane (Betters and Yu, 2010).

Other sterols may also be found naturally in human diet: plant sterols and plant stanols, collectively known as phytosterols (PS), with daily intakes ranging from 150 to 450 mg, depending on dietary choices (Ostlund, 2002). The main sources of plant sterols are vegetable oils, spreads and margarines, cereals, and vegetables, corresponding to 50-80% of the daily intake, with fruits contributing to a further 12%. The most abundant are sitosterol and campesterol, which contribute 60% and 20%, respectively of plant sterol intake. Plant stanols, predominantly sitostanol and campestanol, especially abundant in cereals (wheat and rye), are present in the diet in much lower amounts, about 17-24 mg/day (Pironen and Lampi, 2004).

The two classes of PS, plant sterols and plant stanols, are absorbed differentially by the human body, respectively in the 0.5-1.9% and <0.1% range (Ostlund, 2007). Due to this low fractional absorption and also efficient excretion into bile, circulating levels of PS are low, ranging from 0.3-1.0 mg/dL for plant sterols, and from 0.002-0.012 mg/dL for plant stanols; these levels are of the order of 500-fold and 10000-fold lower, respectively, than those of cholesterol (Bjorkhem et al., 2001).

PS are taken up into tissues in similar proportions to cholesterol: the ratio of cholesterol to PS in tissues is similar to that or less than that in plasma. In healthy individuals, data on tissue levels of PS are available for the carotid artery wall and for non-stenotic and stenotic aortic valve cusps: total plant sterol and plant stanol concentrations are approximately 0.04 μg/mg and 0.001 μg/mg of tissue, respectively, while total cholesterol content is 20 μg/mg of tissue (Miettinen et al., 2005; Helske et al., 2008; Miettinen et al., 2011).

PS are lipophilic steroid alcohols found in plants which are chemically similar to cholesterol, but differ in side chains and ring (cyclopentane-perhydro-phenanthrene) structure saturation. Due to the chemical similarity of PS to cholesterol and to the fact that they are handled by most cells in the same way as cholesterol, PS contribute to the TC and LDL-C lowering, when supplemented in the diet via competitive inhibition of cholesterol absorption (Ostlund, 2007). PS compete with cholesterol for incorporation into the mixed micelles, diminishing its solubilization, and thus reducing cholesterol absorption. PS may also compete with cholesterol during NPC1L1 mediated transport, as well as packaging into chylomicrons in the enterocytes. PS are preferentially transported back into the lumen by ABCG5/8. Once in the circulatory system, PS are also preferentially excreted from the liver into the bile by the same ABCG5/8 heterodimer transporter (Ostlund, 2007). In sum, PS act by lowering bioavailability of intestinal cholesterol for entry into the circulation.

Thus it follows that the cholesterol absorption pathway presents itself as an attractive clinical opportunity for dietary supplementation with cholesterol-lowering agents, such as PS.

## 2. Factors Influencing Phytosterol Supplementation in Human Diet: Dose, Frequency and Time of Intake, Type, Delivery, Formulation, and Individual Modifiers

The ability of PS to lower TC and LDL-C levels was demonstrated for the first time in 1953 in humans (Pollack, 1953), and has been since then supported by many clinical trials, as evidenced in several meta-analyses (Katan et al., 2003; AbuMweis et al., 2008; Demonty et al., 2009; Talati et al., 2010; Musa-Veloso et al., 2011; Genser et al., 2012; and Ras et al., 2013).

Given that the level of PS intake (1 to 3 g/day) required for cholesterol lowering ranges well above the typical consumption patterns, several categories of foods (yellow fat spreads; milk, yoghurt and cheese type products; spicy sauces and salad dressing; and certain rye breads) and nutraceuticals (capsules and tablets) with added free or esterified PS have been globally marketed, since the introduction of functional foods containing PS (specifically margarine supplemented with plant stanol esters) in the Finnish market around 1995. And, on the basis of the available information, PS are effective in all food and capsule/tablet formats (Clifton, 2015). But, the magnitude of LDL-cholesterol reduction may vary depending on the dose, frequency and time of intake, type, delivery, and formulation of the PS-enriched foods supplemented in the diet, to which adds the effect of individual modifiers.

In the end, what matters is that PS are delivered effectively to compete with cholesterol for absorption.

## Dose Effect

Demonty et al. (2009) conducted a meta-analysis and demonstrated that PS lower LDL-C in a dose-dependent manner (2.15 g/day of PS yielded a LDL-C lowering of 0.34 mmol/L or 8.8%). This same conclusion has been noted by others (AbuMweis et al., 2008; Katan et al., 2003; Talati et al., 2010; Musa-Veloso et al., 2011); the 2 g/day dose is the dose currently recommended by several scientific bodies. The European Atherosclerosis Society (EAS) Consensus Panel appraised evidence relevant to the benefit to risk relationship of functional foods with added PS and concluded that 2g PS/day cause significant inhibition of cholesterol absorption and lower LDL-C levels by 8-10%, thereby lower cardiovascular risk (Gylling et al., 2014).

The 3 g/day level is considered to represent the ceiling at which the dose-response curve for cholesterol-lowering with PS begins to plateau, but additional studies are needed to better characterize the LDL-C lowering effects of PS at intakes in excess of 2 g/day (Musa-Veloso et al., 2011).

## Frequency and Time of Intake

Whether the effect of a single dose consumed at different time of the day is comparable to that consumed as multiple dosages throughout the day remains under investigation, but for now PS enriched products are being marketed for consumption once a day, because it is thought that a single dose increases consumers' compliance and adds convenience (AbuMweis et al., 2008).

Additionally, modelling showed that consumption on more than three occasions per day or daily consumption of two or more products, at the recommended intake level, was necessary to exceed a daily intake of 3 g of PS (European Food Safety Authority, 2008).

## Type of PS

Despite large meta-analyses showing that no matter the type of PS supplemented, both plant sterols and plant stanols appear to work equivalently (at daily intakes currently recommended) in improving serum cholesterol levels, the debate comparing the efficacy of both types is still going on, because there's a lack of trials using a head-to-head comparison, looking at higher doses than those currently recommended (Musa-Veloso et al., 2011).

## Delivery

The delivery vehicle is important in predicting successful cholesterol lowering, and currently PS supplementation is available in food (in different types of matrices) or nutraceutical (capsules and tablets) forms.

AbuMweis et al. (2008) investigated in a meta-analysis the effect of several food matrices enriched with PS on LDL-C lowering: greater LDL cholesterol-reduction efficacy was found when PS were incorporated into fat spreads, mayonnaise and salad dressing, and milk and yoghurt in comparison with other food products. Demonty et al. (2009) also compared the effect of high fat to low fat food products and no differences in the absolute or relative dose response curve were seen, although PS-enriched fat foods may not be ideal for individuals following low fat diets to lower cholesterol.

In what concerns the consistency (solid versus liquid) of the delivery matrix on LDL-C lowering, Demonty et al. (2009) reported that at PS intakes greater than 2 g/day, delivery in solid matrices was associated with larger reductions in LDL-C compared to delivery in liquid matrices, possibly due to the longer gastrointestinal transit times of solid foods relative to liquid foods. Musa-Veloso et al. (2011) reported no existence of such effect on LDL-C lowering.

PS-enriched foods and PS nutraceuticals have demonstrated equivalent hypocholesterolemic efficacy. Mackay and Jones (2011) reviewed the cholesterol-lowering action in the limited number of published PS trials using capsules or tablets. These nutraceuticals included plant stanol and sterol esters and also free stanols, in a dose free PS range of 0.8 to 3.0 g/day, for intervention periods of 4 to 12 weeks. Over 50% of the studies reported significant reductions in LDL-C ranging from 4-11%.

Consumers nowadays have a wide choice in the marketplace, in the delivery of PS supplementation, ranging from functional foods to nutraceuticals. Whether as an adjunct to lifestyle or pharmacotherapy, the cost of these products may be of relevance, since the cost of the products with added PS can range 1.3-fold to up 4-fold higher than their counterparts (European Food Safety Authority, 2008). But, in the end, it will be the consumer's preferences that will dictate the adoption of one or another behavior.

## Formulation

Several meta-analysis of clinical trials have found no significant difference between LDL-C lowering response with esterified or free PS-enriched foods (AbuMweis et al., 2008; Demonty et al., 2009; Musa-Veloso et al., 2011). An advantage of free PS is a lower added mass (30% less without the fatty acid) and lower calories (Clifton, 2015). Nevertheless esterification of PS may be important for LDL-C lowering as it facilitates their incorporation into micelles. It also facilitates greater incorporation of PS into foods, and ameliorates the palatability, which enhances compliance and effectiveness. Moreover, incorporation techniques for esterified and free PS are in expansion and therefore offer consumers a wider choice, which may increase its use.

## Inter-Individual Variability

Substantial variability in LDL cholesterol-lowering efficacy exists in responsiveness to PS supplementation from person to person. Nutrigenetic variations across genes associated with cholesterol trafficking pathways are certainly important in predicting the individual response to dietary interventions (Jones, 2015). Within studies, the variability data in terms of efficacy of PS intervention shows that two thirds of individuals fall into a range between zero reduction and twice that of the mean level of cholesterol-lowering (Jones, 2015). The leading meta-analyses have signaled this phenotypic variation in response, but haven't given an explanation of what might cause some individuals to benefit from PS intervention whereas others do not. A possible cause might be related to the reciprocity observed between rates of synthesis and absorption efficiency for cholesterol that exists across individuals. Rideout et al. (2010) showed that subjects with high basal cholesterol synthesis are less responsive to PS treatment than are subjects with low basal cholesterol synthesis.

More, reductions in LDL-C levels are also greater in individuals with high baseline LDL-C levels compared with those with normal to borderline levels.

These phenotypic differences in responsiveness appear to be predicted by genotypic variations in genes coding for key carriers and enzymes in cholesterol metabolic pathways, with preliminary data suggesting a link to possession of a specific haplotype for the ApoE and CYP7A1 genes (De Castro Orós et al., 2011). But further studies are needed to investigate the role of genetics in the LDL-C response to PS.

# 3. Phytosterols as Markers of Intestinal Cholesterol Absorption

Cholesterol absorption and cholesterol synthesis need to be measured for the study of the whole-body cholesterol metabolism in humans. Standard absolute methods were developed and validated years ago, but in today clinical lipid research, noncholesterol sterols (NCS) are being widely used as surrogate markers of cholesterol metabolism: cholesterol biosynthetic precursors, especially squalene, desmosterol and lathosterol, as markers of cholesterol synthesis, and cholesterol metabolite 5α-cholestanol, and plant sterols (mainly sitosterol and campesterol), as markers of cholesterol absorption (Kuksis, 2001). Its simultaneous determination in serum samples is a valuable asset particularly in investigating the reciprocal relationship between cholesterol absorption and synthesis, but it is also a tool for health evaluation during lipid-lowering lifestyle/drug therapy monitoring and a way of diagnosing inherited disorders.

The golden methods are the sterol balance technique to measure cholesterol synthesis and the continuous dual oral or oral-intravenous tracers to determine cholesterol absorption. Other available methods to measure the absolute cholesterol synthesis *in vivo* are based in the fractional conversion of squalene; deuterium incorporation, and mass isotopomer distribution analysis.

The absorption techniques with labeled cholesterol intake over a period of time enable the calculation of the overall absorption efficiency or fractional absorption of dietary cholesterol (percentage absorption of dietary cholesterol).

The cholesterol balance method represents the gold standard for calculating cholesterol synthesis, which is estimated as the difference between cholesterol excretion (fecal neutral sterols and bile acids) and assessed dietary intake, during metabolic steady state (constant plasma cholesterol and fecal cholesterol excretion levels during a period of constant weight). Fecal neutral sterols and bile acids are measured in feces by gas chromatography-flame ionization detection (GC-FID) or gas chromatography-mass spectrometry (GC-MS), and fecal flow is monitored with oral administration of markers such as chromium (III) oxide ($Cr_2O_3$) (Mackay and Jones, 2011).

Although absolute methods have evolved from radioisotopes to stable isotopes, and from invasive to less invasive procedures, these techniques are methodologically complicated and time consuming, especially in large populations.

Advances in analytical methods for the determination of serum NCS have enabled its wider use and validation as surrogate markers of cholesterol metabolism, in several populations and during different interventions.

The analytical methods for the determination of cholesterol absorption and synthesis surrogate markers were reviewed by Mackay & Jones, 2011 and Andrade et al., 2013. Mackay et al. (2014) showed that the ideal analytical method for the measurement of cholesterol and NCS is parallel GC-FID and GC-MS, the only reported to be completely validated.

For example, using these serum cholesterol metabolism markers, large population and cohort studies have been performed to find out whether high cholesterol absorption, high cholesterol synthesis, or individual elevated levels of plant sterols or cholestanol are associated with the risk of coronary artery disease (CAD). But, in the final approach to assess the whole-body cholesterol metabolism, besides the use of these surrogate markers, a verification is also needed with the absolute measurements (Gylling, 2014).

Several trials have investigated the effect on lipids circulating levels and also on serum cholesterol metabolism markers of a variety of PS-enriched food matrices in hypercholesterolemic patients, either alone or in combination treatment, and this number is constantly increasing.

# 4. Phytosterols Health Benefits, Safety, Efficacy and Effectiveness

PS in different formats at doses of 2-3 g per day are effective and safe cholesterol-lowering agents with many clinical applications: adjuncts to a healthy diet, hypercholesterolemia, and combination treatment with statins, metabolic syndrome, and diabetes (Cofán and Ros, 2015).

Therefore, PS-enriched foods despite not being a substitute for a cardioprotective diet in the management of hypercholesterolemia, should be recommended as part of healthy lifestyle, as they are bioactive in reducing cholesterol absorption and blood cholesterol levels.

Besides its role in the non-pharmacological management of hypercholesterolemia in the general population, it is important to know the response to PS as add-on therapy to patients treated with hypocholesterolemic drugs.

Based on LDL-C lowering and the absence of adverse signals, the European Atherosclerosis Society (EAS) Consensus Panel concluded that functional foods with PS may be considered 1) in individuals with high cholesterol levels at intermediate or low global cardiovascular risk who do not qualify for pharmacotherapy, 2) as an adjunct to pharmacologic therapy in high and very high risk patients who do not achieve LDL-C targets on statins or are intolerant to statins, and 3) in adults and children (>6 years) with familial hypercholesterolemia (Gylling et al., 2014).

It can be hypothesized that PS and statins have a synergistic role in cholesterol reduction. A meta-analysis of eight PS + statin trials showed an additional reduction of TC and LDL-C levels of 0.36 mmol/L and 0.34 mmol/L, respectively, when compared with statin monotherapy, and that the reductions induced were equivalent to doubling the dose of statins alone, suggesting its use to minimize side effects associated with high doses of statins (Scholle et al., 2009).

So far, the combination PS + statin studies evaluating the effect on lipids and NCS, included individuals with baseline LDL-C levels between 3.35 mmol/L (130 mg/dL) and 4.90 mmol/L (190 mg/dL). Recently, the additional effect of a low-fat 2 g/day PS-enriched fermented milk on serum cholesterol absorption and synthesis markers and lipid profile was investigated in the elderly on stable statin therapy, with baseline LDL-C < 3.35 mmol/L (Andrade et al., 2015). A therapeutic clinical benefit was observed, corresponding to a significant LDL cholesterol reduction of 0.15 mmol/L and 0.27 mmol/L from baseline, respectively after 3 and 6 weeks of intervention.

Overall, there is not enough information on the efficacy of PS as add-on therapy to other hypocholesterolemic single-target agents, such as ezetimibe, fibrates or bile acid binding resins (Cofán and Ros, 2015).

The recent IMPROVE-IT trial showed that ezetimibe effectively lowered the number of new CVD events by lowering serum LDL-C concentrations via inhibiting intestinal cholesterol absorption (Kohno, 2014). Although PS and ezetimibe both act by reducing intestinal cholesterol absorption (by comparable mechanisms), these same effects on CVD outcomes have not so far been proven for PS. Yet, the possible advantage of PS over ezetimibe is that these compounds act on multiple targets, extending potential health effects beyond their effects on LDL-C, once they also lower serum triacylglycerol (TAG) concentrations in individuals with high levels of TAG, and might also influence the functioning of the human immune system via a changed activity of the regulatory T-cells, as proposed by Plat et al. (2015).

Because PS effectively reduce LDL-C concentrations, their use could constitute a potential protective strategy to delay or prevent the onset of atherosclerosis. But, most of the supportive data relate to studies on experimental atherosclerosis in animal and cell-based models. Despite its inherent limitations, such effects were correlated with cholesterol lowering based on PS action, and included reduction in arterial lipid accumulation, inhibition of lesion formation/progression, regression of existing lesions, and influence on inflammatory pathways (Gylling et al., 2014).

All the above facts show that PS are non-systemic, non-pharmacological agents for the treatment of hypercholesterolemia. In addition, in 2008 Jones and Varady suggested that PS as functional foods components could be considered essential nutrients. From a clinical perspective, evidence for nutrient essentiality is of importance, and the notion of essentiality has been broadened to afford a level of protection against chronic degenerative diseases. Jew et al. (2015) investigated PS among other bioactives case studies, and have presented evidence for the re-evaluation of PS as essential nutrients, for specific populations and/or physiological conditions where body cholesterol pools are high.

Still, regular intakes of PS with enriched foods, do naturally result in increased plasma PS concentrations (by about a factor of 2) (Silbernagel et al., 2015). This has raised potential health concerns of atherogenicity, although a recent meta-analysis by Genser et al. (2012) did not support such a positive association. Posteriorly, the results of the meta-analysis by Ras et al. (2013) revealed that although a positive association exists between the absolute changes in plasma PS (sitosterol and campesterol) concentrations and the dose of PS, baseline PS concentration and PS composition of the enriched foods, the total PS remain below 1% of total sterols circulating in the blood. According to the recent consensus statement of the EAS, the concerns raised by some scientists were not supported by conclusive evidence (Gylling et al., 2014).

However, higher plasma levels of PS associated with intakes of 2 g/day have not been linked to the adverse effects on health in long-term human studies.

Even though there are no cardiovascular outcome trials for PS-enriched foods, there is strong rationale for using these functional foods to reduce cholesterol and, thereby, reduce cardiovascular risk in hypercholesterolemic populations.

## Conclusion and Final Remarks

Abundant evidence over past decades shows that foods with PS lower serum LDL-C concentrations. The 2 g PS/day dose is currently recognized by authoritative bodies to induce a significant inhibition of cholesterol absorption and lower atherogenic LDL-C levels by an average 8-10%. Present data are as yet unclear about the long-term safety of doses higher than 3 g/day. On the basis of consistent scientific evidence on the cholesterol-lowering efficacy and safety of foods supplemented with PS, several international clinical societies have endorsed their use as adjuncts to a healthy diet. In the setting of statin treatment, the LDL-C reduction with PS combined with statins is equivalent to up titrating twice the statin dose, enhancing the potential for attainment of LDL-C goals as a function of overall cardiovascular risk. However, direct evidence is lacking on CVD outcome benefits after PS (2 g/day) long-term use.

Either to assess the whole-body cholesterol metabolism, or in lipid-lowering diet/drug therapy monitoring, the determination of non-cholesterol sterols in serum samples as surrogate markers of cholesterol metabolism, which includes plant sterols as absorption markers and biosynthetic cholesterol precursors as synthesis markers, has been gaining a wider use in many clinical trials.

## References

AbuMweis, SS; Barake, R; Jones, PJH. Plant sterols/stanols as cholesterol lowering agents: a meta-analysis of randomized controlled trials. *Food Nutr Res*, 2008, 52 doi:10.3402/fnr.v52i0.1811.

Andrade, I; Santos, L; Ramos, F. Advances in analytical methods to study cholesterol metabolism: the determination of serum noncholesterol sterols. *Biomed Chromatogr*, 2013, 27(10), 1234-42. doi:10.1002/bmc.2840.

Andrade, I; Santos, L; Ramos, F. "Add-on" effect of phytosterols-enriched fermented milk on lipids and markers of cholesterol metabolism in statin-treated elderly patients. Steroids, 2015, 99, 293-298. doi:10.1016/j.steroids.2015.03.007

Betters, JL; Yu, L. NPC1L1 and cholesterol transport. *FEBS Lett*, 2010, 584, 2740-2747.

Bjorkhem, I; Boberg, KM; Leitersdorf, E. Inborn errors in bile acid biosynthesis and storage of sterols other than cholesterol. In: Scriver CR; Beaudet al.; Sly WS; Valle D: editors. The metabolic and molecular basis of inherited disease. 8th ed. New York: McGraw-Hill: 2001, 2961-2988.

Clifton, P. Influence of food matrix on sterol and stanol activity. *J AOAC Int*, 2015, 98(3), 1-2.

Cofán, M; Ros, E. Clinical application of plant sterol and stanol products. *J AOAC Int*, 2015, 98(3), 1-6.

De Castro-Orós, I; Pampín, S; Cofán, M; Mozas, P; Pintó, X; Salas.Salvadó, J; Rodriguez-Rey, JC; Ros, E; Civeira, F; Pocoví, M. Promoter variant -204A > C of the cholesterol 7α-hydroxylase gene: association with response to plant sterols in humans and increased transcriptional activity in transfected HepG2 cells. *Clin Nutr*, 2011, 30(2), 239-246. doi:10.1016/j.clnu.2010.07.020.

Demonty, I; Ras, RT; van der Knaap, HC; Duchateau, GS et al. Continuous dose-response relationship of the LDL-cholesterol lowering effect of phytosterol intake. *J Nutr*, 2009, 139, 271-284.

European Food Safety Authority. A report from the data collection and exposure unit in response to a request from the European Commission. Consumption of food and beverages with added plant sterols in the European Union. EFSA J 2008, 133, 1-21. http://www.efsa.europa.eu/en/sdocs/doc/datex_report_ej133_phytosterols_en.pdf [accessed 12.05.2015].

Genser, B; Silbernagel, D; De Backer, G; et al. Plant sterols and cardiovascular disease: a systematic review and meta-analysis. *Eur Heart J*, 2012, 33, 444-51.

Gylling, H. Clinical utility of serum markers of cholesterol absorption and synthesis. *Curr Opin Lipidol*, 2014, 25, 207-12.

Gylling, H; Plat, J; Turley, S; Ginsberg, HN; Ellegard, L; Jessup, W; Jones, PJ; Lütjohann, D; et al. Plant sterols and plant stanols in the management of dyslipidaemia and prevention of cardiovascular disease. *Atherosclerosis*, 2014, 232(2), 346–360.

Helske, S; Miettinen, T; Gylling, H; et al. Accumulation of cholesterol precursors and plat sterols in human stenotic aortic valves. *J Lipid Res*, 2008, 49, 1511-8.

Jew, S; Antoine, J-M; Boulioux, P; Milner, J; Tapsell, LC; Yang, Y; Jones, PJH. Nutrient essentiality revisited. *J Funct Foods*, 2015, 14, 203-209.

Jones, PJH. Inter-individual variability in response to plant sterol and stanol consumption. *J AOAC Int*, 2015, 98(3), 1-5.

Jones, PJ; Varady, KA. Are functional foods redefining nutritional requirements? *Applied Physiol Nutr Metab*, 2008, 33(1), 118-123.

Katan, MB; Grundy, SM; Jones, P; Law, M; et al. Efficacy and safety of plant stanols and sterols in the management of blood cholesterol levels. *Mayo Clin Proc*, 2003, 78, 965-978.

Kohno, T. Report of the American Heart Association (AHA) Scientific Sessions 2014, IL, Circ J 2014, Dec 15 Epub ahead of print.

Kuksis, A. Plasma non-cholesterol sterols. *J Chromat A*, 2001, 935(1-2), 203-236.

Mackay, D; Jones, PJH. Evaluation of methods for the determination of cholesterol absorption and synthesis in humans. *Atherosclerosis*, 2011, 218, 253-62.

Mackay, DS; Jones, PJH; Myrie, SB; Plat, J; Lutjohann, D. Mehodological considerations for the harmonization of non-cholesterol sterol bio-analysis. *J Chromatogr B*, 2014, 957, 116-22.

Miettinen, TA; Railo, M; Lepantalo, M; Gylling, H. Plant sterols in serum and in atherosclerotic plaques of patients undergoing carotid endarterectomy. *J Am Coll Cardiol*, 2005, 45, 1792-801.

Miettinen, TA; Nissinen, M; Lepantalo, M; et al. Non-cholesterol sterols in serum and endarterectomized carotid arteries after a short-term plant stanol and sterol ester challenge. *Nutr Metab Cardiovasc Dis*, 2011, 21, 182-8.

Musa-Veloso, K; Poon, TH; Elliot, JA; Chung, C. A comparison of the LDL-cholesterol lowering efficacy of plant stanols and plant sterols over a continuous dose range: results of a meta-analysis of randomized, placebo-controlled trials. *Prostaglandins Leukot Essent Fatty Acids*, 2011, 85, 9-28.

Ostlund, RE; Jr. Phytosterols in human nutrition. *Ann Rev Nutr*, 2002, 22, 533-549.

Ostlund, RE; Jr. Phytosterols, cholesterol absorption and healthy diets. *Lipids*, 2007, 42, 41-45.

Pironen, V; Lampi, AM. Occurrence and levels of phytosterols in foods. In: Dutta PC, editor. Phytosterols as functional food components and nutraceuticals. New York: Marcel Dekker, Inc: 2004 1-32.

Plat, J; Baumgartner, S; Mensik, P. Mechanisms underlying the health benefits of plant sterol and stanol ester consumption. J AOAC Int 2015, 98(3),1-4.

Pollack, OJ. Reduction of blood cholesterol in man. *Circulation*, 1953, 7, 702-706.

Ras, RT; Hiemstra, H; Lin, Y; Vermeer, MA; Duchateau, GSMJE; Trautwein EA. Consumption of plant sterol-enriched foods and effects on plasma plat sterol concentrations-a meta-analysis of randomized controlled studies. *Atherosclerosis*, 2013, 230, 336-346.

Reiner, Z; Catapano, AL; De Backer, G; Graham, I; Taskinen, M-R; Wiklund, O; Agewall, S; Alegria, E; Chapman, MJ; Durrington, P; Erdine, S; Halcox, J; Hobbs, R; Kjekshus, J; Filardi, PP; Riccardi, G; Storey, RF; Wood, D. (2011). ESC/EAS Guidelines for the management of dyslipidaemias. The Task Force for the management of dyslipidaemias of the European Society of Cardiology (ESC) and the European Atherosclerosis Society (EAS). *Eur Heart J, 32*(14), 1769-1818. http://dx.doi.org/10.1093/eurheartj/ehr158.

Rideout, TC; Harding, SV; Mackay, D; AbuMweis, SS; Jones, PJ. High basal fractional cholesterol synthesis is associated with nonresponse of plasma LDL cholesterol to plant sterol therapy. *Am J Clin Nutr*, 2010, 92, 41-46. doi10.3945/ajcn.2009.29073.

Scholle, JM; Baker, WL; Talati, R; Coleman, CI. The effect of adding plant sterols or stanols to statin therapy in hypercholesterolemic patients: systematic review and meta-analysis. *J Am Coll Nutr*, 2009, 28(5), 517-524.

Silbernagel, G; Baumgartner, I; Marz, W. Cardiovascular Safety of plant sterol and stanol consumption. *J AOAC Int*, 2015, 98(3), 1-3.

Talati, R; Sobieraj, DM; Makanji, SS; Phung, OJ; Coleman, CI. The comparative efficacy of plant sterols and stanols on serum lipids: a systematic review and meta-analysis. *J Am Diet Assoc*, 2010, 110, 719-726.

Temel, RE; Brown, JM. Biliary and nonbiliary contributions to reverse cholesterol transport. *Curr Opin Lipidol*, 2012, 23, 85-90.

Turner, S; Voogt, J; Davidson, M; et al. Measurement of reverse cholesterol transport pathways in humans: in vivo rates of free cholesterol efflux: esterification, and excretion. *J Am Heart Assoc*, 2012, 1, e001826.

In: Phytosterols
Editor: Deanna Garner

ISBN: 978-1-63483-477-3


*Chapter 6*

# THE PRACTICAL USE OF SOME NON-CHOLESTEROL STEROLS (LATHOSTEROL, DESMOSTEROL, CAMPESTEROL AND SITOSTEROL) IN DIFFERENTIATION AND TREATMENT OF CHILDREN AND ADOLESCENTS SUFFERING FROM HYPERCHOLESTEROLEMIAS

***Josef Hyanek*[1,*], *Frantisek Pehal*[2], *Ladislava Dubska*[2], *Vera Martinikova*[1], *Jana Privarova*[1] and *Ludek Taborsky*[2]**
[1]Lipid Outpatient Clinic
[2]Department of Clinical Biochemistry, Hematology and Immunology,
Na Homolce Hospital, Prague, Czech Republic
(Nemocnice Na Homolce, Prague, Czech Republic

## SUMMARY

***Background:*** The article presents selected plasmatic cholesterol precursors (Lathosterol – Lat, Desmosterol – Des) and plasmatic phytosterols (Campesterol—Cam, Sitosterol—Sit) in relation to the parameters of cholesterol metabolism in 300 children patients (8-18 yrs) with genetically proved heterozygous familial hypercholesterolemias (HFH) and other hypercholesterolemias in order to obtain practical experience with the possible differential and monitoring use of these parameters in dietary or drug treatment. The authors discuss their 20 years of experience with the use of these seleceted non-cholesterol sterols (NCS) in diagnosing, differentiation, and dietary & drug treatment.

***Methods:*** The plasma NCS values were determined using GC/MS; plasma lipid profile was established by routine clinical biochemistry methods; and molecular genetic analyses were conducted by PCR-SSCP and PCR-DGGE methods.

***Results:*** The article indicates reference values of NCS for genetically proved healthy children of different ages. Patients with the so-called 'alimentary hypercholesterolemia'

[*] Corresponding author: Josef Hyanek, tel.+420 257 273 229; e-mail: josef.hyanek@homolka.cz.

possess normal values of NCS. Children patients with HFH detected by positive selective screening (5th and 13th year of life) and arriving at the lipid outpatient clinic had usually been put on a low cholesterol diet (LCD) by pediatricians and had elevated levels of Lat and Des and compensatorily elevated Cam and Sit. Administration of simvastatin in sufficient therapeutic doses reduced Lat or Des ($p > 0.001$). Compensatory elevation of Cam and Sit persists in a few patients. Ezetimibe application decreases efficiently Cam and Sit rather than Lat and Des. Combined application of simvastatin with ezetimibe is the most efficient not only in the decrease in the plasma LDL-cholesterol, but also in decrease in Lat, Des as well as Cam and Sit. The diagnostic and therapeutic monitoring in patients with different types of hypercholesterolemia is also given in the graphic form.

***Conclusions:*** Extending the laboratory monitoring profile by precursors of cholesterol synthesis (Lat, Des) and cholesterol absorption (Cam, Sit) improves the differential diagnosis power for different hypercholesterolemias in children and makes monitoring and treatment of children and adolescents suffering from HFH more precise and effective.

**Keywords:** non-cholesterol sterols, lathosterol, desmosterol, campesterol, sitosterol, familial hypercholesterolemia, dietary and drug treatment, statins, ezetimibe

## 1. Introduction

Cholesterol is the most important zoosterol. It is essential for each and every human cell. Its highly variable homeostasis is dependent on its synthesis, absorption and excretion, because the total cholesterol level in the human plasma is a result of at least 5 regulatory and genetically independent mechanisms regulating the cholesterol balance and homeostasis: (a) clearance of LDL particles by the cholesterol receptors in the liver cells; (b) endogenous synthesis in the adrenal glands, liver, intestines, etc.; (c) absorption of non-cholesterol sterols in the intestine; (d) enterohepatic circulation of bile acids between the liver and in the small intestine; (e) pregnancy or secondary disease and its pathophysiological effects, such as thyroid diseases, diabetes, etc. To understand such a mass of regulatory mechanisms and provide for effective dietary and drug treatment, it is necessary to employ a wide range of tested metabolites, which is broader than the classical lipid profile used for many years, i.e., total cholesterol (TC); Low-Density Lipoprotein Cholesterol (LDL); High-Density Lipoprotein Cholesterol (HDL) and triacylglycerols (TAG). Non-cholesterol sterols (NCS) clearly demonstrate how the range of the effective markers can be broadened to improve diagnostics, monitoring and treatment. Thanks to the chromatographic analytical methods of the last twenty years, cholesterol synthesis and cholesterol absorption can be quantitatively measured in serum by suitable selected NCS.

Practising prevention and treatment of hereditary disorders of lipid metabolism in children on a daily basis for over twenty years, we have chosen (because of their easy methodological availability) and gathered experience with selected markers which cover both cholesterol synthesis (e.g., lathosterol—Lat, and desmosterol—Des) and absorption (e. g. campesterol—Cam, and sitosterol—Sit). Thanks to the efforts of the Finnish researchers around Prof. Miettinen, vast experience has been collected not only with the metabolites mentioned above, but also with many others, such as squalene, which is not affected by statins, as well as cholestanol, which has been repeatedly used to monitor the intestine absorption, etc. There is a lot of knowledge available about NCS, especially about the negative correlation between the cholesterol synthesis markers and phytosterol absorption markers, but available practical findings and experience about hypercholesterolemia diagnostics and treatment are scarse for adults and none as regards children [1, 2, 3, 4, 5, 6].

Patients suffering from inherited hypercholesterolemias are at a high risk of premature cardio- or cerebrovascular disorders (CVD). Early identification of these inherited diseases allows administration of proper lifestyle and drug therapy as early as in childhood to prevent cardiovascular or cerebrovascular complications [7-10].

## 2. Classification, Differentiation and Preventive Screening of Cholesterol Disorders

Among the many forms of hypercholesterolemia, those characterized by marked elevation of TC and LDL are considered the most important. They are: (1) hereditary familial hypercholesterolemias (HFH) resulting from deficient LDL receptor proteins that cannot (i) bind LDL (binding defects), (ii) internalize LDL (internalizing defects), (iii) disrupt the normal recycling of the LDL (recycling defects); and (2) hereditary familial ligand defect of apoB-100 (FDB) hypercholesterolemia.

It is generally known that HFH with autosomal dominant mode of inheritance (OMIM 143890) cannot be successfully treated by even a strict low-cholesterol diet (LCD), and requires differentiated drug therapy. Recently, it has been finally accepted that pediatric HFH (boys from the age of 10 years and girls after first menstruation) require the same intensity of medical therapy as adults [11-18].

But what are our options for early distinction between the most severe forms of HFH and the so-called “alimentary hypercholesterolemia” (AH), which is caused by overeating and an unhealthy lifestyle and forms the majority of cases encountered in our lipid clinic? Since the clinical symptoms of HFH in children have not yet been developed as sufficiently as in adults, we must rely on the combination of the typical laboratory findings based on the lipid profile, and the molecular genetic examination of LDL receptors. Therefore, we have been including selected noncholesterol sterols (NCS), specifically cholesterol synthesis precursors and cholesterol absorption precursors, in the routine testing for over 20 years.

Following the introduction of obligatory nationwide selective screening for hyperlipoproteinemia in the Czech Republic for children from high-risk families in their fifth and thirteenth year of life (since 1994), those attending our metabolic lipid clinic for further examination are not only children of the patients operated on in our hospital with a surgically documented risk of CVD, but chiefly children with a range of different hypercholesterolemia symptoms identified in pediatric primary care settings [19]. These symptoms are subsequently differentiated and the required therapy is determined depending on the results—either LCD in cases of AH or drug treatment in cases of HFH.

Encouraged by the pioneering work of Hamilton, Miettinen, Kempen, Jones, and Gylling [3, 4, 21, 22, 23, 24, 25, 26], as well as the promising latest research, which offers a growing mass of surprising knowledge on polymorphisms in the ATP-binding cassette transporters G5, G8, A1, we combine the standard lipid marker examination with the most easily available markers of cholesterol synthesis precursors—lathosterol (Lat) and desmosterol (Des)—, and the cholesterol absorption precursors—campesterol (Cam) and beta-sitosterol (Sit)—for better hypercholesterolemia differentiation [1, 2, 5, 6, 27, 28, 29, 30].

The most important structures of NCS are shown in Figure 1. For simplified metabolic schemes of cholesterol synthesis and absorption processes see Figure 2 and Figure 3, respectively.

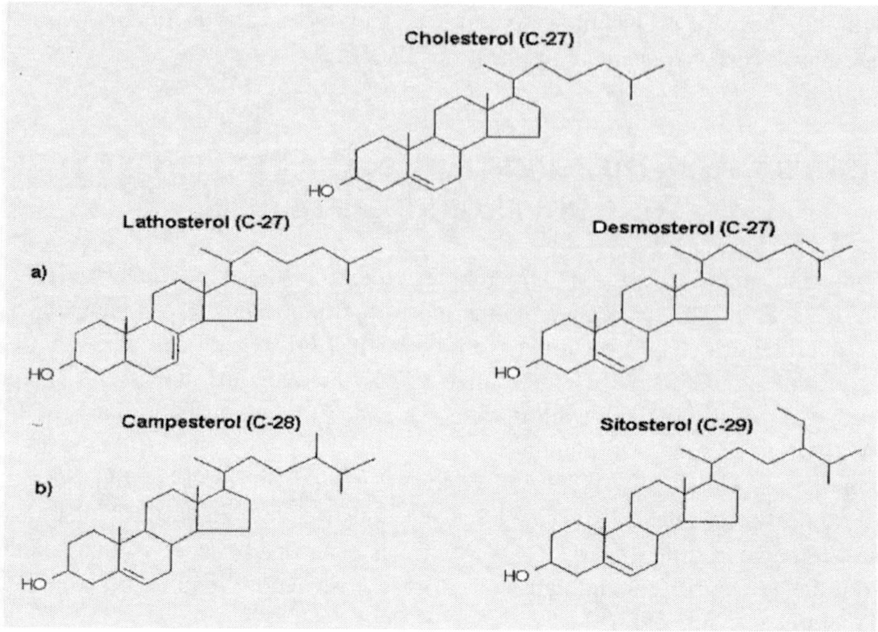


Figure 1. Structures of cholesterol, and the diagnosticaly most important phytosterols working as its (a) synthesis and (b) absorption precursors.

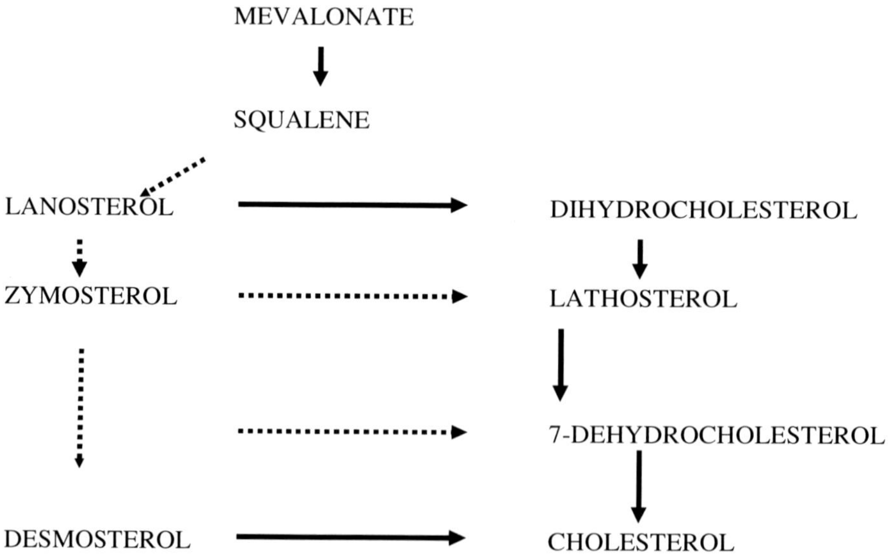


Figure 2. Simplified metabolism of cholesterol synthesis in man.

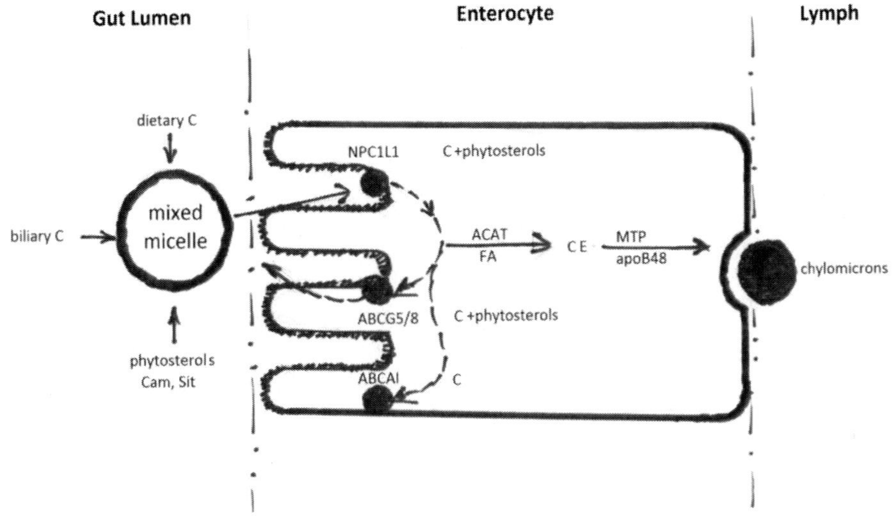


Figure 3. Simplified pathways of absorption of cholesterol and phytosterols in human enterocyte (based on [53]).

## 3. Patients Definition and Methods

- 108 statin and ezetimibe treated children (10-18 yrs) with a positive family history of CVD (Framingham Risk Score 10-20%) and the genetically proved deficit of LDL-receptor mutations, specifically in p. Gly592Glu (17 patients), p. Cys200Tyr (6 patients), p. Pro424_Asn425ins32 (5 patients), p. Gly218del (5 patients), p. Arg4+6Trp (3 patients), p. Asp266Glu (4 patients), p. Gly396Alafsx54 (5 patients), p. Ser177Leu (3 patients), p. Trp 666X (5 patients), p. His690ThrfsX19 (3 patients), p. Cys143Arg (3 patients), p. Asp227Glu (3 patients), p. Gly149Cys (2 patients), p. Gly176Val (4 patients), p. Ala299Val (2 patients), Ala 232Asn27+del (2 patients); and with the familial ligand-defective apoB 100 (FDB) 3500W mutation (2 homozygote patients and 31 heterozygote patients);
- Another 156 statin, ezetimibe or LCD treated children of the same age with metabolic findings resembling HFH, where molecular genetic analysis has not yet confirmed LDL-reception deficiency (familial combined hyperlipidemias and polygenic hypercholesterolemias);
- 84 healthy school-age children and adolescents (6-18 years old) from schools in Prague area as a control group;
- 37 children of the same age with confirmed so-called "alimentary hypercholesterolemia," with high cholesterolemia frequently accompanied by hypertriacylglycerolemia (>3 mmol/l), hyperglycaemia and mild obesity;
- 35 randomly selected healthy blood donors (18-60 years old) as a control group for NCS in healthy adults.

The basic characteristics of the groups investigated are shown in Table 1.

The LCD was defined as 1992 NCEP-Ped Step II diet (<300 mg/d cholesterol in natural food) [33]. The "Alimentary Hypercholesterolemia" was determined on the basis of a lipid profile assessment, pedigree analysis, low risk of CVD in the family, and evaluation of a submitted two-week food journal, which was analyzed for nutrient content and cholesterol intake (mg/d) by the Ostrasoft Co. (1992) software programme [34]. According to recommendations by Rifkind, deJong, and Wiegman [11, 14, 35], "individual cholesterol tolerance" was respected so as for TC blood levels not to increase over 4.8 mmol/l.

Medical treatment involves the use of simvastatin—Zocor (MSD) in individual doses of 5-20 mg/d—, which has been successfully combined with ezetimibe—Ezetrol (MSD) in individual doses of 5-10 mg/d—over the past 12 years [11, 14, 17, 18]. During the two-month summer vacations, when the patients are usualy not available for monitoring and drug therapy, the drug treatment is often insufficiently observed, and the only remaining treatment is LCD.

Monitoring of the diet and drug treatment is ensured by biochemical and immunological testing of lipid markers at 3-6 month intervals (depending on the patient's compliance); levels of NCS, lipid soluble vitamins, sex hormones and trace elements in 6-month intervals. Height and weight are recorded using a software programme, and secondary signs of sex maturation are registered according to Tanner and Goldstein [36] on each visit to the clinic. Performed annually, skinfold measurements are taken by the Best caliper, and IMT is measured using a VigmedSound device.

The routine biochemical and classical lipid profile (TC colorimetric assay, HDL immunoassay, LDL direct immunoassay, TAG enzymatically) is determined using a Synchron Beckman Coulter LX 20; haematology profile by means of an Advia 120 Bayer; apo Al and apo B nephelometrically using an Immage; apo E polymorphism is measured by molecular genetic analysis; sexual hormone levels are determined with the IRMA method using a Stratec SR300; trace elements using a GT-AAS Varian; lipid soluble vitamins by means of HPLC with UV detection; vitamin $D_3$ with the RIA technique using a Multicrystal Berthold LB211; vitamin B6 using a Chromsystem; holotranscobalamin using an Abbot AXIM; and beta-carotene by means of HPLC.

The following DNA analyses of genes for LDL-receptors and FDB were performed: direct DNA analysis, sequential analysis of exon 4 and 5, and analysis of deletions and duplications of the LDL-receptor gene and the entire region using the Multiplex Ligation Probe Amplification method, carried out through the goodwill of the Genetic Laboratory, Centre of Cardiovascular and Transplantation Surgery, Brno (MedPed project Czech National Coordinator, MUDr. T. Freiberger, PhD.) [37]. PCSK9 genes were not sequenced. The first degree relatives of HFH probands with identified gene mutations underwent DNA sequencing regardless of the LDL plasma levels.

Plasma NCS were determined first by GC following Phillips et al. [38], later using the Theunissen [39] modification by means of a GC/MS Finnigan Mat 1020B: internal standard epicoprostenal, CV < 10%. The values of detectable lanosterol are very low without any possible diagnostic use. According to our practical experience and a longitudinal evaluation of the NCS levels, Des essentially follows in lower concentrations than Lat values, and Sit follows Cam values in a similar way: we have retained only these two markers in routine testing over the recent years and found them to be sufficient to register both cholesterol synthesis and phytosterol absorption.

**Table 1. Basic metabolic and demographic data of investigated groups of healthy children, healthy adults and child patients suffering from hypercholesterolemias (median values)**

| | Age (yrs) | Women/ Men | Skin Fold (mm) | BMI | TC (mmol/l) | ApoB (g/l) |
|---|---|---|---|---|---|---|
| **68 Children on LCD** | 10 to 18 | 43/25 | 5.6/10.3/8.9 | | 6.7 +/- 1.5 | 1.86 |
| **108 Children on statins** | 10 to 18 | 58/50 | 5.5/10.2/8.9 | | 5.0 +/- 0.5 | 1.45 |
| **84 Healthy children** | 6 to 18 | 52/32 | 4.2/8.5/7.5 | | 3.7 +/- 1.3 | 1.12 |
| **35 Healthy adults** | 20 to 65 | 12/13 | | 28.1 | 5.2 +/- 1.6 | 1.62 |
| **37 Alimentary hypercholesterolemias** | 6 to 18 | 18/19 | 5.8/10.8/9.2 | | 6.6 +/- 2.5 | 1.02 |

The overall lipid extraction from the serum samples was performed using the Folch method [50]. First, the inner standard of epicholesterol (5 μg) was added to the serum sample (0.9 ml). Methanol (0.9 ml) was added, and the whole mixture was agitated for 15 minutes. The extraction was done using chloroform (2 x 0.9 ml) while adding water solution of NaCl. The mixture was then centrifuged (at 200 g) for 10 minutes.

The lipid saponification was performed using the modified Thompson and Merola method [51]. Following evaporation of chloroform ethanol with 3% pyrogalol (8 ml) was added to the lipid extract sample. The preparation was briefly mixed by hand, and KOH (0.5 ml, 1.28g/ml) solution was added. The saponification (30 min. at 88°C) was performed following mixing and sonification of each sample. Cyclohexan (20 ml) and distilled water (12 ml) were added to each sample and the content was agitated. After centrifugation (at 200 g) the upper layer of cyclohexan was transferred into a vial, and evaporated under a nitrogen stream (approx. 65°C).

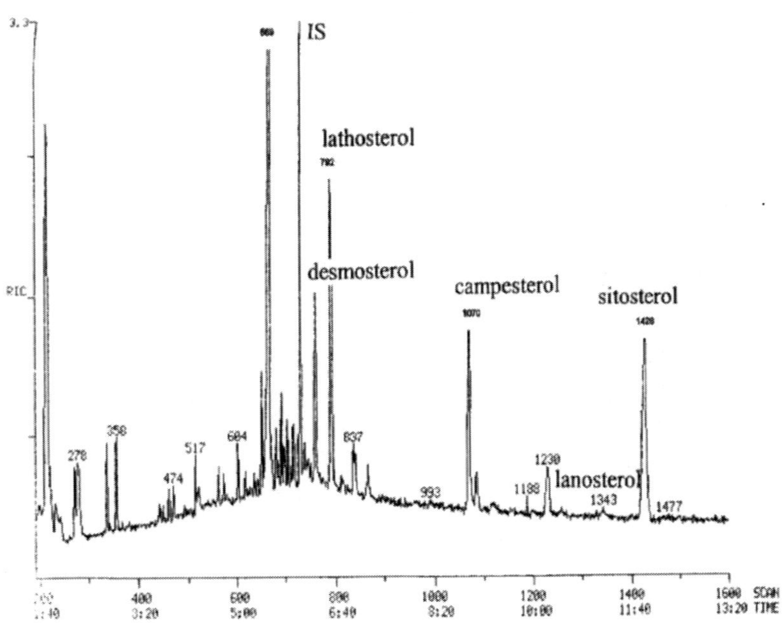


Figure 4. Elution spectrum of analyzed non-cholesterol sterols.

Another step in the preparation of samples for GC/MS was their extraction derivatization using methanol-methyl chloroformate-pyridine. The reason was derivatization of sterols while separating the neutral lipid fraction from polar lipids. The derivatization was performed using the modified Hušek method [45]. The extracted compounds were then analysed on the Focus GC with ITQ 700 ion trap mass spectrometer (manufactured by Thermo Electron Corporation) heated to 200°C. TR-1 column (30 m length, 0.25 mm internal diameter, 0.25 µm film of dimethyl polysiloxane) was used for separation. The ion source was operated in the electron ionisation mode (70 eV, 20 µA) and the mass spectra were obtained using the ion trap operating in a SIM scan mode.

The elution spectrum of the most important NCS is shown in Figure 4.

For expression of NCS serum values in this paper absolute values are used in µmol/l, as well as in µmol/l divided by total cholesterol in mmol/l x 100 as ratios, which according to Miettinen and Kempen better express the true NCS balance. [6, 12, 13].

The study was approved by the Na Homolce Hospital Ethical Committee as part of Project No. NA 7452-3 supported by the Internal Grant Agency of the Czech Ministry of Health (IGA MZ CR).

## 4. Statistical Analysis

Statistical analysis was performed using MedCalc Version 8.1.1.1 (Belgium). The basic statistical parameters included mean or median calculation (95% CI), number of elements in sample, maximum and minimum values, and relevant standard deviations. The normal distribution of the set was tested using the D'Agostino-Pearson test. For graphical comparison of patient groups Box-and-Whisker plots were used with the median and 25-75 percentil.

Paired t-test was used to compare the two independent means of NCS measured values and lipid metabolites during treatment. The difference was significant with values $<0.05$.

## 5. Results

The basic metabolic and demographic characteristics of the individual groups of children and adolescents under survey are given in Tab. 1. A pilot study in healthy school-age children had to be conducted first in order to obtain basic data on NCS among the Czech child population. The results for the group of 84 healthy school-age children used as reference values are given in Table 2. The values for the healthy subjects are dominated by Cam, followed by Sit, while Lat and Des levels are lower. Despite the slightly lower levels in the youngest children, there is not much variance in the overall values between the healthy children and healthy adults.

Children with HFH who have been on longitudinal LCD, with which they often arrive at the outpatient lipid clinic, have always higher levels of Lat and Des (due to higher synthesis) and of Cam and Sit only if parents use the margarine enriched food (Tab. 3, Figure 5a, 5b, 5c, 5d and Figure 8-10).

Cam and Sit are also increased because the parents usually enrich the LCD with plant sterol margarines following the pediatrician's recommendation (Table 3.).

In the group of children with documented LDL-R mutations, the increase in Lat and Des is usually greater than in the group of children with hypercholesterolemia where LDL-receptor deficiency has not been confirmed (Figures 5a, 5b).

Children with AH attending for the first time the metabolic unit usually do not have elevated levels of Lat and Des compared to children with HFH (Table 4, Figures 5a, 5b).

**Table 2. Reference values of non-cholesterol sterols in 84 healthy children and 25 healthy adults from Czech population (μmol/l)**

| | Lathosterol | Desmosterol | Campesterol | Sitosterol |
|---|---|---|---|---|
| **6-8yrs** | | | | |
| **girls** | 4.0 +/- 0.8 | 2.6 +/- 0.5 | 9.9 +/- 1.2 | 7.4 +/- 0.9 |
| **boys** | 3.7 +/- 0.9 | 2.2 +/- 0.6 | 9.7 +/- 1.4 | 7.5 +/- 1.1 |
| **8-12yrs** | | | | |
| **girls** | 6.0 +/- 0.6 | 3.2 +/- 0.4 | 11.0 +/- 1.0 | 7.7 +/- 0.7 |
| **boys** | 6.0 +/- 0.6 | 3.3 +/- 0.4 | 9.7 +/- 0.9 | 7.2 +/- 0.7 |
| **12-15yrs** | | | | |
| **girls** | 7.2 +/- 0.5 | 3.8 +/- 0.3 | 10.5 +/- 0.7 | 7.0 +/- 0.5 |
| **boys** | 6.1 +/- 0.5 | 3.5 +/- 0.3 | 9.6 +/- 0.8 | 7.0 +/- 0.6 |
| **15-18yrs** | | | | |
| **girls** | 5.7 +/- 0.8 | 3.1 +/- 0.5 | 10.1 +/- 1.2 | 5.8 +/- 0.9 |
| **boys** | 6.3 +/- 1.3 | 3.3 +/- 0.8 | 16.0 +/- 2.0 | 7.4 +/- 1.5 |
| **adults 18-65yrs** | | | | |
| **women** | 6.4 +/- 0.7 | 4.2 +/- 0.4 | 10.7 +/- 1.0 | 8.7 +/- 0.8 |
| **men** | 7.4 +/- 1.1 | 3.9 +/- 0.6 | 9.1 +/- 1.6 | 7.7 +/- 1.2 |

**Table 3. Average plasmatic levels of non-cholesterol sterols in healthy children compared with low-cholesterol diet treated patiens suffering from heterozygous familial hypercholesterolemia**

| | | Lat | Des | Cam | Sit |
|---|---|---|---|---|---|
| | **number** | **μmol/l** | **μmol/l** | **μmol/l** | **μmol/l** |
| | | **Healthy children** | | | |
| **girls** | 52 | 4.8 +/- 0.4 | 2.8 +/- 0.2 | 9.3 +/- 0.7 | 6.2 +/- 0.5 |
| **boys** | 32 | 4.6 +/- 0.5 | 2.4 +/- 0.3 | 9.4 +/- 0.9 | 6.4 +/- 0.6 |
| | | **Children on low-cholesterol diet (300 mg/d)** | | | |
| **girls** | 35 | 12.2 +/- 1.2 | 5.3 +/- 0.7 | 11.7 +/- 1.9 | 7.3 +/- 1.4 |
| **boys** | 33 | 8.6 +/- 1.2 | 4.5 +/- 0.6 | 11.6 +/- 1.7 | 9.2 +/- 1.2 |
| **t-test** | | p < 0.0001 | p < 0.0001 | p < 0.0001 | p < 0.0001 |

**Table 4. Non-cholesterol sterols in (a) 37 children with "alimentary hypercholesterolemia" and (b) 38 children with heterozygous familial hypercholesterolemia**

| NCS | (a) Median (+/- SD) | (b) Median (+/- SD) | Significance (p<) |
|---|---|---|---|
| Lat (µmol/l) | 4.6 +/- 0.93 | 8.95 +/- 2.7 | 0.0001 |
| Lat/TCx100 | 92.0 +/- 18.6 | 129 +/- 63.2 | 0.0011 |
| Cam (µmol/l) | 8.6 +/- 2.5 | 7.7 +/- 5.1 | 0.3786 |
| Cam/TCx100 | 168 +/- 47.5 | 117 +/- 103 | 0.0211 |
| Des (µmol/l) | 2.9 +/- 0.89 | 4.9 +/- 1.69 | 0.0011 |
| Des/TCx100 | 92+/-18 | 59+/-22 | 0.1469 |
| Sit (µmol/l) | 6.3 +/- 2.4 | 6.0 +/- 3.6 | 0.6898 |
| Sit/TCx100 | 104+/-49 | 74+/-45 | 0.025 |

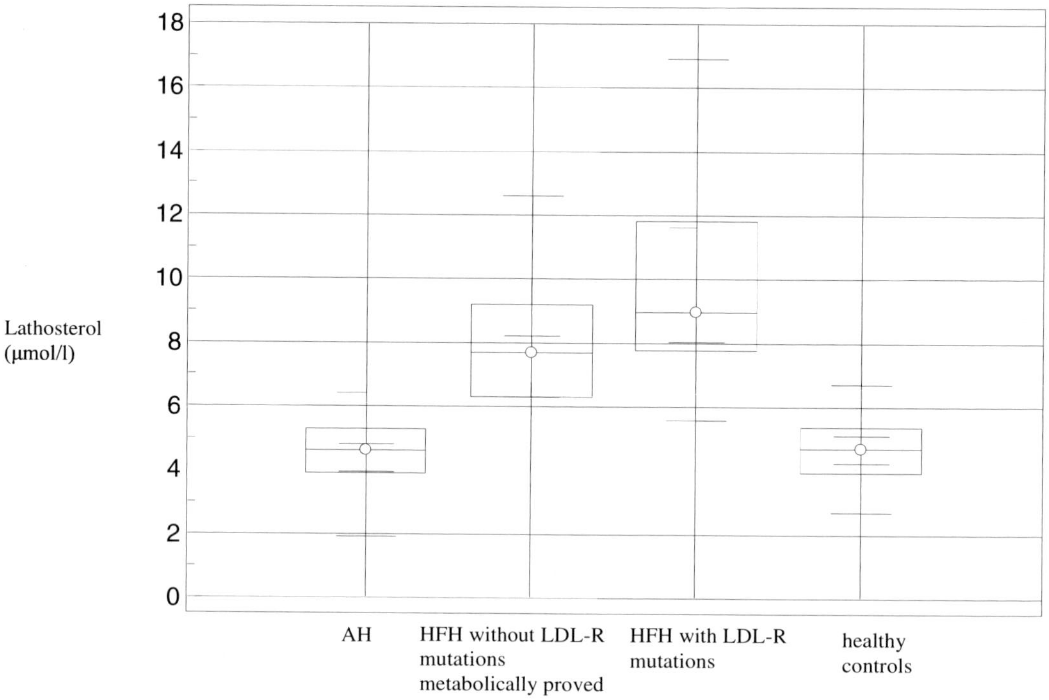


Figure 5a. Box plot of lathosterol concentrations in different hypercholesterolemias (before treatment)—box and whisker medians (error bars: 95% CI for median).

Statins administrated in effective doses reduce TC as well as Lat levels. Alterations of both in absolute values correlate with the TC lowering response ($p < 0.001$). It is only when statin doses are too low that the Lat level persists or falls slowly (Figure 6).

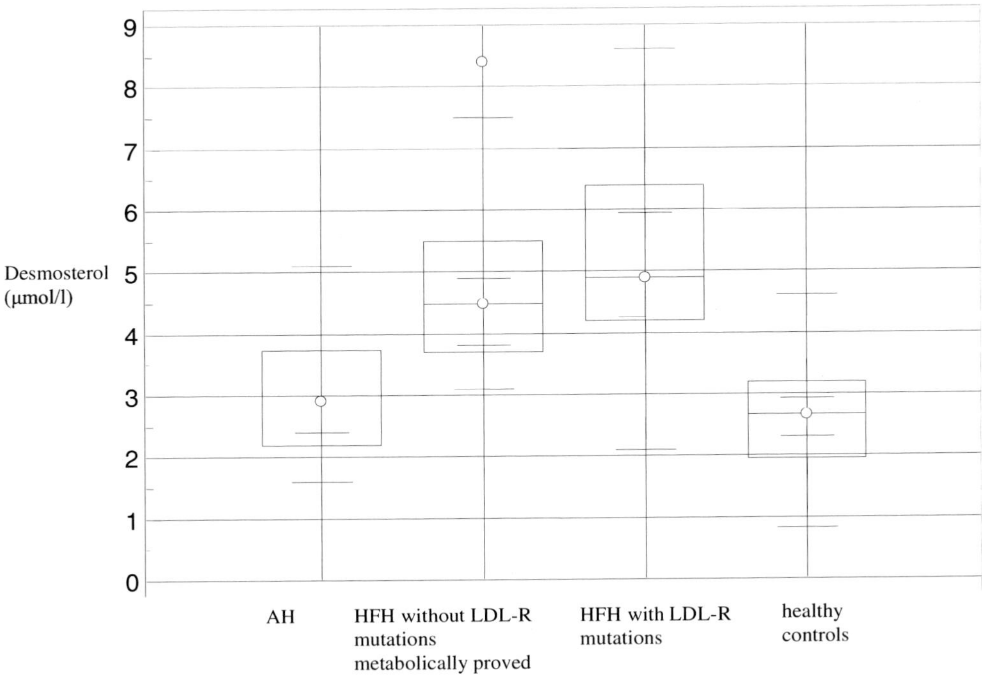


Figure 5b. Box plot of desmosterol concentrations in different hypercholesterolemias (before treatment)—box and whisker medians (error bars: 95% CI for median).

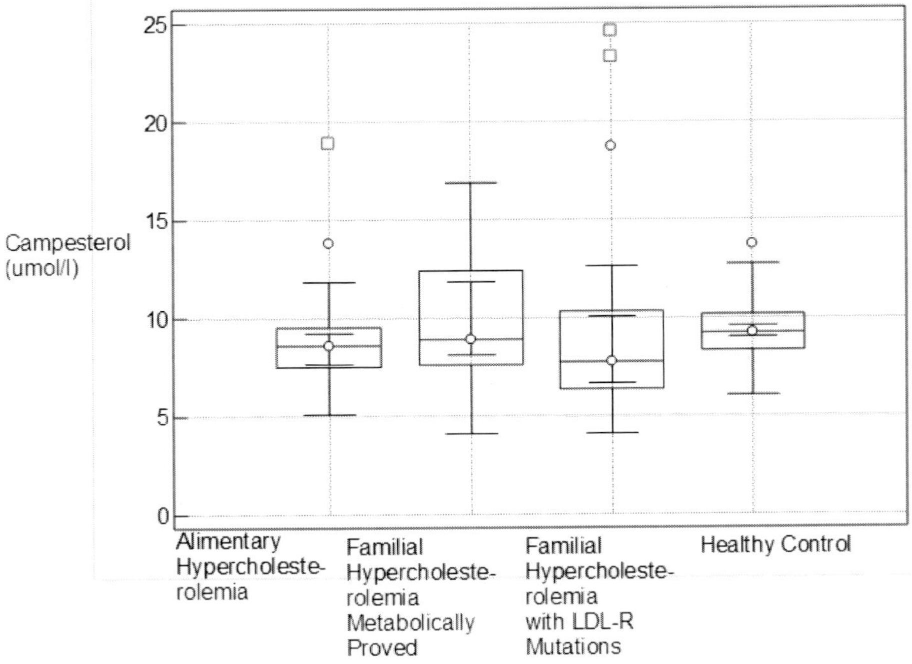


Figure 5c. Box plot of campesterol concentrations in different hypercholesterolemias (before treatment)—box and whisker medians (error bars: 95% CI for median).

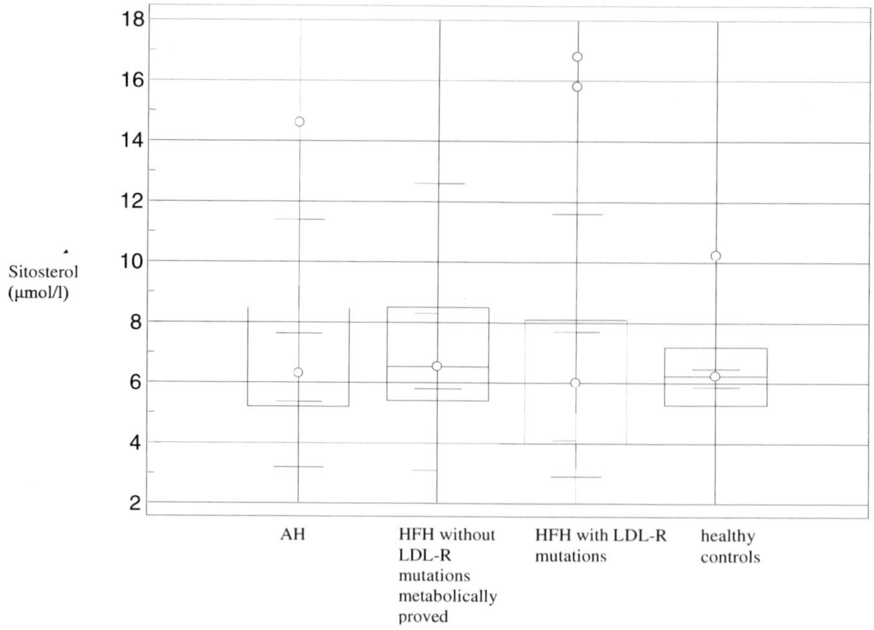


Figure 5d. Box plot of sitosterol concentrations in different hypercholesterolemias (before treatment)—box and whisker medians (error bars: 95% CI for median).

Doses of statin can be individually set for each patient according to the intensity of its Lat levels decrease. Compensatory increase of Cam and Sit has only been observed in a few patiens. Combined treatment of statin with ezetimibe has to date produced the most effective decrease in both Lat and Des along with decreased TC, while the Cam or Sit are not always reduced (Figures 7a,7b, 7c).

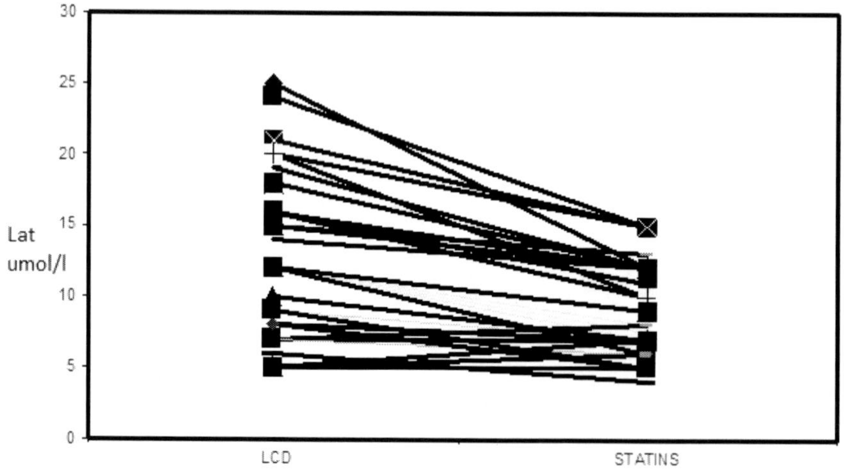


Figure 6. Lathosterol decrease after interruption of low-cholesterol diet and introduction of statin therapy in 35 children patients with heterozygous familial hypercholesterolemia.

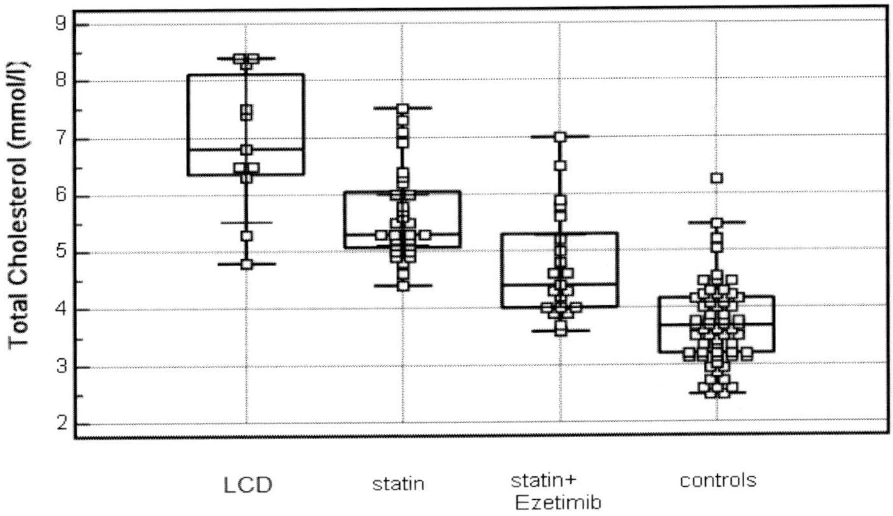


Figure 7a. Box plot of total cholesterol concentrations in hypercholesterolemias according to the type of dietary or drug treatment—box and whisker medians (error bars: 95% CI for median).

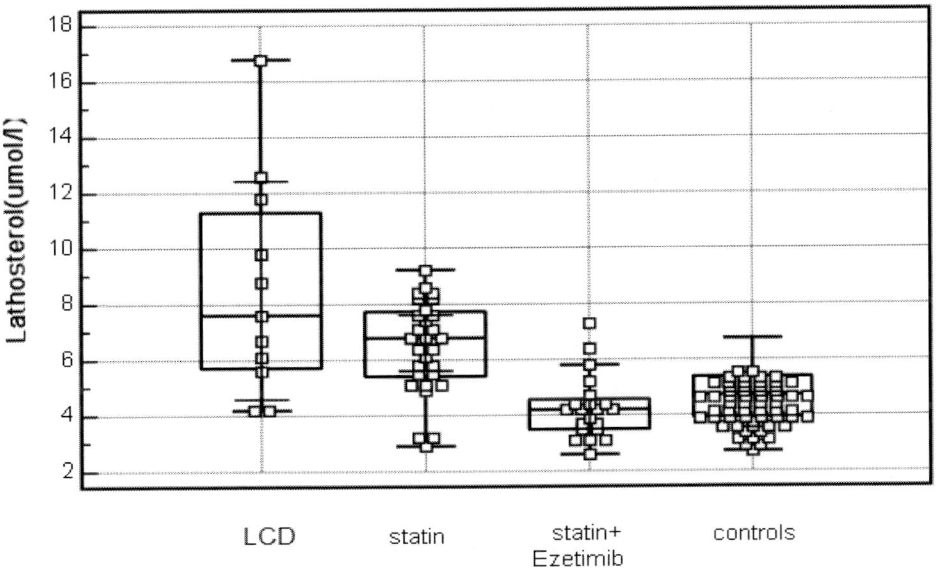


Figure 7b. Box plot of lathosterol concentrations in hypercholesterolemias according to the type of dietary or drug treatment—box and whisker medians (error bars: 95% CI for median).

Another interesting oscillation of Lat has been observed when we carefully monitored lathosterol in 30 child patients with HFH also during vacations. We could observe the "typical increase" in plasmatic Lat levels where the LCD was applied instead of the targeted drug therapy. Our simple explanation was that during vacations the children are often without proper parental control for drug treatment, and therefore only the LCD is usually guaranteed, which results in immediate Lat level increase.

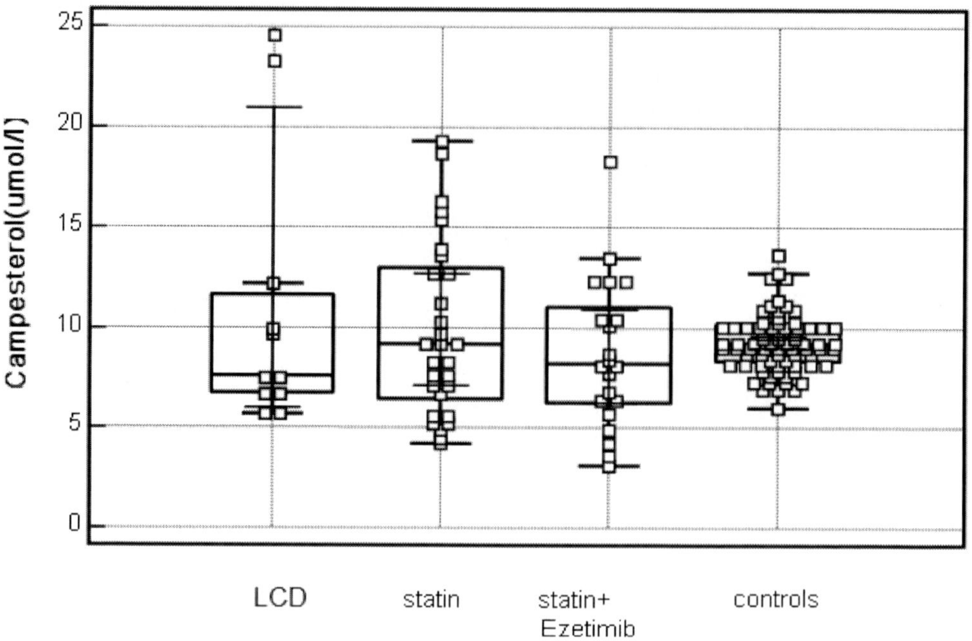


Figure 7c. Box plot of campesterol concentrations in hypercholesterolemias according to the type of dietary or drug treatment—box and whisker medians (error bars: 95% CI for median).

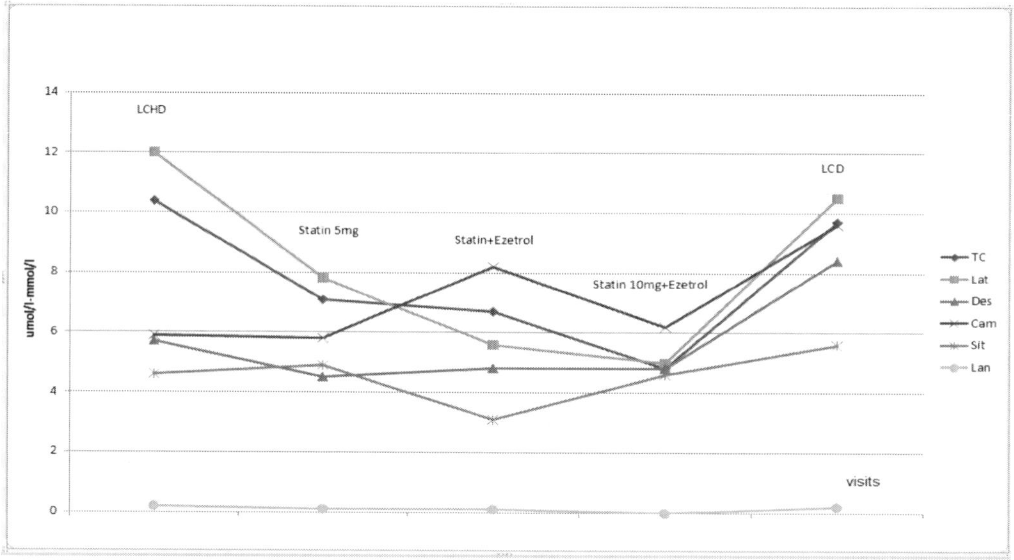


Figure 8. Monitoring of non-cholesterol sterol plasmatic levels in a homozygous patient with the familial ligand-defective apoB 100 mutation during dietary and drug treatment.

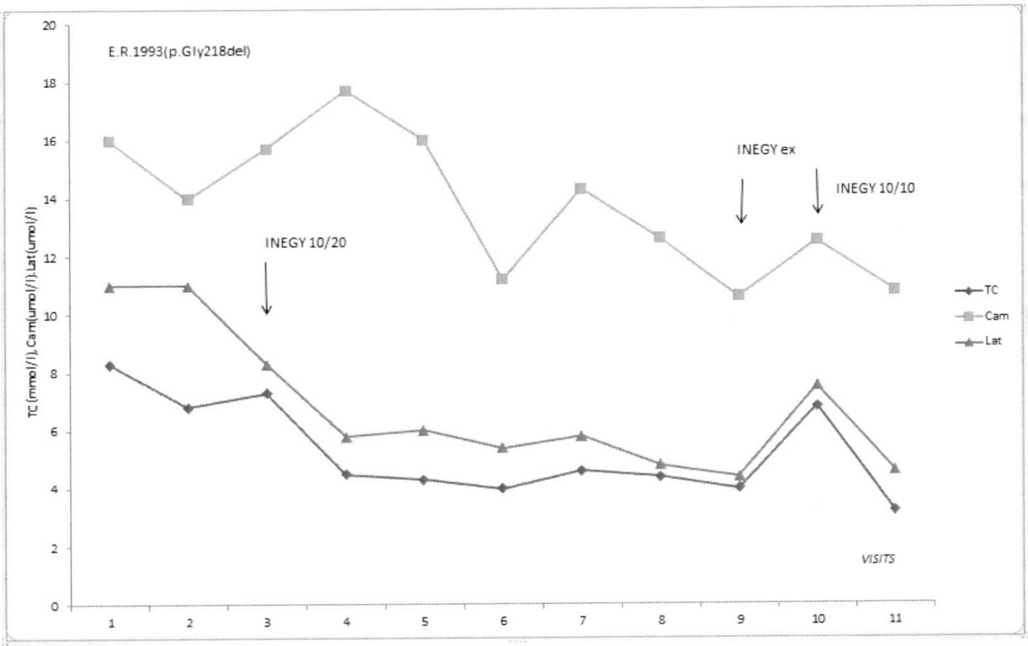


Figure 9. Monitoring of lathosterol and campesterol levels in a heterozygous patient with the p. Gly592Glu LDL-receptor mutation during dietary and drug treatment.

Monitoring Lat and other NCS during dietary and drug treatment in a 18-year homozygous female patient with FDB is shown in Figure 8. The following figures (Figures 9 & 10) show the Lat and Cam levels in two heterozygous patients with different forms of HFH.

## 6. DISCUSSION

Every physician working in a lipid clinic certainly has experience with the failure of LCD in the treatment of HFH, even when the parents and children comply with this treatment [11, 12, 13, 14, 15, 54]. We have all also encountered strict mothers or grandmothers who apply even stricter LCD (200mg/d) than necessary (leading more often than not to protein deficiency) in their attempt to protect their children and grandchildren from the increased risk of vascular failure, which they have experienced among their closest relatives. Such attempts often result in reduced growth as indicated by the children's growth curves. On the other hand, we also see the easily gullible mothers, and especially grandmothers, accompanying typically "overfed children" with TC and TAG levels of 8-12 mmol/l, who only follow television advertising or clamorous leaflets offering "guaranteed" dietary products from domestic and foreign food companies. We have therefore enlarged the range of markers for the differential diagnosis of hypercholesterolemias in order to simplify the system as much as possible and avoid harming the children.

A lot has been publicized about metabolism, physiological and pathobiochemical findings of NCS, esp. in connection with CVD, but little research is still available on the practical experience with its daily use [28, 30, 32, 54]. In our decision making as regards treatment in our clinic we have been so far most helped by Lat levels. We have patients where Lat decrease correlates with TC decrease to the therapeutical level. Cam levels are usually increased at the beginning of treatment in patients on LCD enriched with plant sterol esters. The compensatory increase of Cam and Sit in statin therapy as postulated by Miettinen, Descamps, Sudhop, Santosa, Hedman and al. [1, 26, 28, 29, 40, 41] have been observed only in a few patients. Our explanation is that the doses of statin used in combined therapy with ezetimibe in our children patients were lower than in adults, mostly below 20 mg/d (Figures7-10). Some children cannot be clearly classified as belonging to any of the above mentioned groups based upon their first metabolic evaluation, and they maintain enormously high Lat values or high Cam values during various types of treatment. However, we do agree with the recommendation of Miettinen et al. that effective statin therapy is mainly observed in the "synthetizer" type patients with high levels of Lat and Des at admission, whereas ezetimibe treatment is more effective in the "absorber" type patients with higher Cam and Sit levels [1, 5, 26, 28, 29, 30]. The combination of ezetimibe and statin is the most effective and least intrusive form of medical treatment for HFH, which fully corresponds with Sudhop's observations [43].

In some patients it is necessary to probably respect the types of patients described a long time ago by Katan et al. as "hyperresponders or hyporesponders" [42]. LCD continues to be reserved primarily for AH, as well as for patients with mixed dyslipidemia, where no molecular genetic examination is yet available or the lipid laboratory profile does not allow for successful differentiation [13].

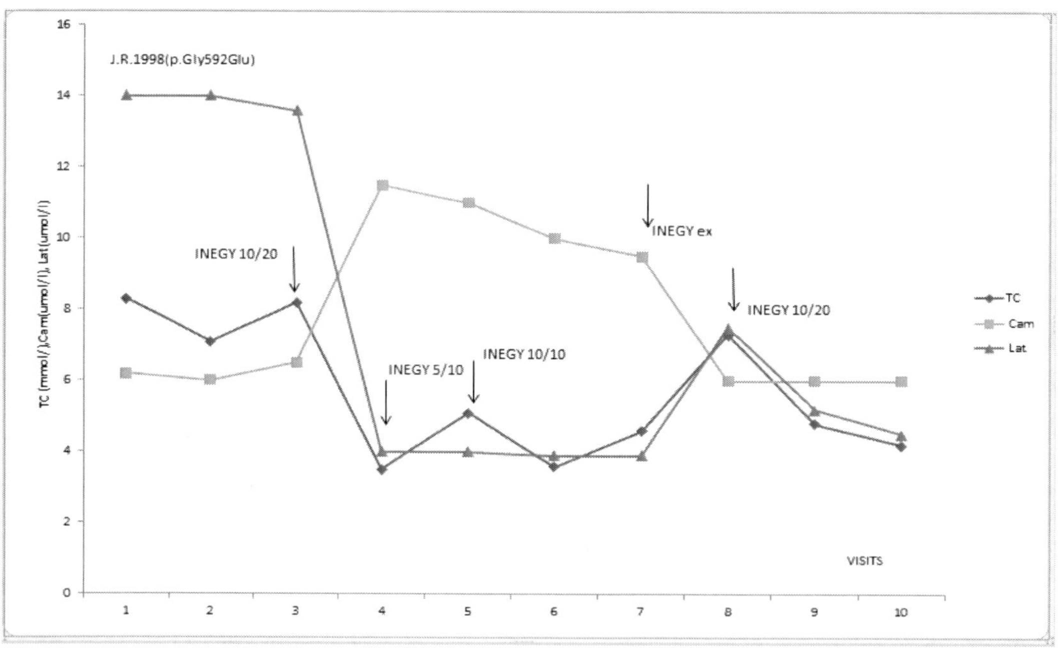


Figure 10. Monitoring of lathosterol and campesterol levels in a heterozygous patient with the p. Gly218del LDL-receptor mutation during dietary and medical treatment.

The possible correlation between low cholesterol synthesis of Lat and high absorption of Cam as seen by Weingartner and other authors in patiens proving a high risk of CVD could not be assessed in our children patiens with HFH [26, 44, 45, 46, 54].

It should also be noted in this place that phytosterols are also atherogenic substances just like cholesterol, and their increased intake is just as harmful to our health as high levels of cholesterol [43, 47].

Given the existence of around 1,000 identified mutations in LDL-receptors with different functional capacities for cholesterol internalization (from 5 to 70%), the yet not fully understandable sophisticated polymorfisms in ATP binding casette transporters G5, G8 or A1, and also given the proven influence of other atherosclerotic risk factors, such as high Lp (a) >1000 mg/l, or adverse polymorphism in apo E 3/4 and 4/4, it is not at all surprising to find such a wide range of hypercholesterolemic patients with diverse reactions to any type of treatment. Therefore, it is necessary to enlarge the methodology, compare practical experience, standardize the testing and verify the observations of different authors so that the discovered values of NCS can really be of practical daily use to clinic physicians, as observed and recommended by MacKay, Thompson et al., Noto et al., and Hedman et al. [31, 32, 46, 48, 49].

To us, the evaluation of NCS in the lipid outpatient clinic is of practical use:

- in the differentiation of heterozygous familial hypercholesterolemia and alimentary hypercholesterolemia;
- in determining the size of therapeutic statin doses;
- in monitoring effective medication (monotherapy or combined therapy);
- for checking whether the patient is actually taking the recommended medication and for verification of the actual intake of food enriched with plant sterol esters.

## ACKNOWLEDGMENTS

Dr. Hyanek reports receiving research support from Merck Sharp & Doehme and Thermo Fischer Scientific.

## ABBREVIATIONS USED

| | | |
|---|---|---|
| ABCG5/8 | — | ATP-binding casette transporter G5/G8 |
| ABCA1 | — | ATP-binding casette transporter A1 |
| ACAT | — | Acetyl-CoA cholesterol acyltransferase |
| ApoB | — | apolipoprotein B |
| AH | — | alimentary hypercholesterolemia |
| C | — | cholesterol |
| Cam | — | campesterol |
| CE | — | cholesteryl esters |
| CV | — | coefficient of variation |
| CVD | — | cardiovascular disorders |

| | | |
|---|---|---|
| Des | — | desmosterol |
| FA | — | fatty acids |
| FDB | — | familial ligand- defective apoB 100 |
| GS | — | gas chromatography |
| GC/MS | — | gas chromatography/mass spectrometry |
| HDL | — | high-density lipoprotein cholesterol |
| HFH | — | heterozygous familial hypercholesterolemia |
| HPLC | — | high-pressure liquid chromatography |
| IMT | — | intima media thickness |
| Lat | — | lathosterol |
| LCD | — | low-cholesterol diet |
| LDL | — | low density lipoprotein cholesterol |
| LDL-R | — | LDL-receptor |
| Lp(a) | — | lipoprotein (a) |
| MPT | — | metabolic transport protein |
| MSD | — | Merck Sharp & Dohme |
| NCEP | — | National Cholesterol Education Program (USA) |
| NCS | — | non-cholesterol sterols |
| NPC1L1 | — | Nieman PickC1Like1 protein |
| OMIM | — | Online Mendelian Inheritance in Man |
| PCR-SSCP | — | polymerase chain reaction—single-strand conformation polymorphism |
| PCR-DGGE | — | polymerase chain reaction denaturing gradient gel electrophoresis |
| RIA | — | raidoimmunoasssay |
| Sit | — | sitosterol |
| TAG | — | triacylglycerols |
| TC | — | total cholesterol |

## References

[1] Miettinen, TA; Gylling, H. Synthesis and absorption markers of cholesterol in serum and lipoproteins during a large dose of statin treatment. *Eur. J. Clin. Invest*, 2003, 33, 967-982.

[2] Miettinen, TA; Gylling, H. Effect of statins on noncholessterol sterol levels: implications for use of plant stanols and sterols. *Am. J. Cardiol*, 2005, 96 (suppl): 40D-46D.

[3] Gylling, H. Cholesterol metabolism and its implications for therapautic interventions in patients with hypercholesterolemia. *Int. J. Clin. Practice*, 2004, 58, 859-866.

[4] Gylling, H; Miettinen, TA. The effect of plant stanol- and sterol-enriched foods on lipid metabolism, serum lipids and coronary heart disease. *Ann. Clin. Biochem.*, 2005, 42, 254-263.

[5] Miettinen, TA; Gylling, H; Viikari, J; et al. Synthesis and absorption of cholesterol in Finnish boys by serum non-cholesterol sterols. The cardiovascular risk in young Finns study. *Atherosclerosis*, 2008, 200, 177-183.

[6] Miettinen, TA; Gylling, H; Nissinen, MJ. The role of serum non-cholestertol sterols as surrogate markers of absolute cholesterol synthesis and absorption. Nutrition, *Metabolism and Cardiovascular Diseases*, 2011, 21, 765-769.

[7] Expert panel on integrated guidelines for cardiovascular health and risk reduction in children and adolescents. *Pediatrics*, 2011, S 213-S256.

[8] Magnussen, CG; Venn, A; Thomson, R; et al. The association of pediatric low-density lipoprotein cholesterol and high-density lipoprotein cholesterol dyslipidemia classifications and change in dyslipidemia satus with carotid intimal-media thickness in adulthood evidence from the cardiovascular risk in Young Finns Study, the Bogalusa Heart Study, and the CDAH (Childhood Determinants of Adult Health) Study. *J. Am. Coll. Cardiol*, 2009, 53, 860-869.

[9] McCrindle, BW; Urbina, EM; Dennison, BA; et al. Drug therapy of high-risk lipid abnormalities in children and adolescents: a specific statement from the American Heart Association Atherosclerosis, Hypertension and Obesity in Youth Committee, Council of Cardiovascular Disease in the Young, with the Council on Cardiovascular Nursing. *Circulation*, 2007, 115, 1948-67.

[10] Gidding, SS; Dennison BA; Birch LL; et al. Dietary recommendations for children and adolescents: a guide for practitioners: consensus statement from the American Heart Association. *Circulation*, 2005, 112, 2061-75.

[11] deJongh, S; Ose, L; Szamosi. T; Gagne, C; Lambert, M. et al. Efficacy and safety of statin therapy in children with familial hypercholesterolemia: a randomized, double-blind, placebo controlled trial with simvastatin. *Circulation*, 2002, 106, 2231-37.

[12] deJongh, S; Vissers, MN; Rol, P; et al. Plant sterol lower LDL cholesterol without improving endothelial function in prepubertal children with familial hypercholesterolemia. *J. Inh. Metab. Dis*, 2003, 26, 343-351.

[13] Obarzanek, E; Kimm, SYS; Barton, BA; et al. Long term safety and efficacy of a cholesterol-lowering diet in children with elevated low-density lipoprotein cholesterol: Seven year results of the Dietary Intervention Study in Children (DISC). *Pediatrics*, 2011, 107, 256-64.

[14] Wiegman, A; Hutten, BA; deGroot, E; et al. Efficacy and safety of statin therapy in children with familial hypercholesterolemia: a randomized controlled trial. *JAMA*, 2004, 292, 331-337.

[15] Rodenburg J; Vissers, MN; Wiegman, Av; Trotsenburg VP; et al: Statin treatment in children with familial hypercholesterolemia. The younger, the better. *Circulation*, 2007, 116, 664-668.

[16] Avis, HJ; Hutten, BA; Gagné, V; et al. Efficacy and safety of rosuvastatin therapy for children with familial hypercholesterolemia. *J. Am. Coll. Cardiol*, 2010, 55, 1121-26.

[17] Gauer, R. Hyperlipidemia treatment in children:The younger, the better. *Amer. Family Physician*, 2010, 82, 462-477.

[18] Clauss, S; Wai, KM; Kavey, RW; Kuehl, K. Ezetimibe treatment of pediatric patients with hypercholesterolemia. *J. Pediat*, 2009, 154, 869-872.

[19] Health Regulations—Preventive Screenings. Bulletin of the Czech Ministry of Health, 7/1994 (in Czech).

[20] Miettinen, TA. Cholesterol precursors and their diurnal rhythm in lipoproteins of patients with jejunoileal bypass and ileal dysfunction. *Metabolism*, 1985, 34, 425-30.

[21] Gylling, H; Miettinen, TA. Serum noncholesterol sterols related to cholesterol metabolism in familial hypercholesterolemia. *Clin. Chim. Acta*, 1988, 178, 41-50.

[22] Kempen, HJM; Glaatz, JFC; Leuven JAG; et al. Serum lathosterol concentration is an indicator of whole-body cholesterol synthesis in humans. *J. Lipid. Res*, 1988, 29, 1149-55.

[23] Miettinen, TA; Tilovbis, RS; Kesaniemi, YA. Serum plant sterols and cholesterol precursors reflect cholesterol absorption and synthesis in volonteers of a randomly selected male population. *Am. J. Epidemiol*, 1990, 131, 20-31.

[24] Jones, PJ; Schoelleer, DA. Evidence for diurnal periodicity in human cholesterol synthesis. *J. Lipid. Res*, 1990, 312, 667-73.

[25] Hamilton, JJ; Hany, M; Innis, SM. Elevation of plasma lathosterol, as indicator of increased cholesterol synthesis in preterm infants given Intralipid. *Pediat. Res*, 1992, 31, 186-192.

[26] Miettinen, TA; Gylling, H; Strandberg, TE; et al. for Finish 4S Investigators. Baseline serum cholestanol as predictor of recurrent coronary events in subgroups of Scandinavian Simvastatin Survival Study. *Brit. Med. J*, 1998, 316, 1127-30.

[27] Strandberg, TE; Tilvis, RS; Pitkala, KH; et al. Cholessterol and glucose metabolism and recurrent cartdiovascular events among elderly. *J. Amer. Coll. Cardiol*, 2006, 48, 708-14.

[28] Hedman, M; Miettinen, TA; Gylling, H; Ketomäki, A; Antikainen, M. Serum noncholesterol sterols in children with heterozygous familial hypercholesterolemia undergoing pravastatin therapy. *J. Pediat*, 2006, 148, 241-246.

[29] Santosa, S; Varady, KA, Abumweis, S; Jones. P. Physiological and therapeutic factors affecting cholesterol metabolism; Does a reciprocal relationship between cholesterol absorption and synthesis really exist ? *Life Sci*, 2007, 80, 505-514.

[30] Nissien, MJ; Miettinen, TE; Gylling, H; Miettinen, TA. Applicability of noncholesterol sterols in predicting response in cholesterol metabolism to simvastatin and fluvastatin treatment among hypercholesterolemic men. *Nutr. Metab. Cardiovasc. Dis*, 2010, 20, 308-316.

[31] Assman, GK; Kannenberg, DR; Ramey, TA; et al. Effects of ezetimibe, simvastatin, atorvastatin, and ezetimibe-statin therapies on non-cholesterol sterols in patients with primary hypercholesterolemia. *Curr. Med. Res. Opin*, 2008, 24, 249-259.

[32] Noto, D; Cefalu, ABC; Barraco, G; et al. Plasma non-cholesterol sterols: a useful diagnostic tool in pediatric hypercholesterolemia. *Pediatr. Res*, 2010, 67, 200-204.

[33] NCEP Expert panel of blood cholesterol levels in children and Adolescents. National Cholesterol Education Program (NCEP): Highlights of the Report of the Expert Panel on Blood Cholesterol Levels in Children and Adolescents. Pediatrics, 1992, 89, 495-501. American Academy of Pediatrics. Statement on Cholesterol. Committee on Nutrition. Pediatrics, 1992, 90, 469-473. The population approach: nutrition recommendations for healthy children and adolescents. *Pediatrics*, 1992, 89, 537-544.

[34] Hyánek, J; Martiníková, V; Maťoška, V; et al. 10-year experience with medical treatment of familial hypercholesterolemia in children and adolescents. Česko-slov. *pediatrie*, 2006, 61, 404-412 (in Czech with English summary).

[35] Rifkind, BM; Segal, P. Lipid research clinic program reference values for hyperlipidemia and hypolipidemia. *JAMA*, 1983, 250, 1869-1872.

[36] Tanner, JM. Growth and Adolescence, 2nd ed. Oxford, Blackwell UK, 1962.

[37] Francova, H; Trbusek, M; Zapletalova, P; et al. New promoter mutations in the low-density lipoprotein receptor gene which induce familial hypercholesterolemia phenotype. Molecular and functional analysis. *J. Inher. Metab. Dis*, 2004, 27, 523-528.

[38] Phillips, KM; Ruggio, DM; Bailey, JA. Precise quantitative determination of phytosterols, stenols, and cholesterol metabolites in human serum, by capillary gas-liquid chromatography. *J. Chromatog*, 1999, B732, 17-29.

[39] Theunissen, CE; Regente, J; Bergman, K; et al. Serum cholesterol precursors and metabolites and cognitive performace in an aging population. *Neurobiol. Aging*, 2003, 24, 147-155.

[40] Descamps, OS; DeSutter, J; Guillaume, M; Missault, L. Where does the interplay between cholesterol absorption and synthesis in the context of statin and/or ezetimibe treatment stand today? *Atherosclerosis*, 217, 2011, 308-321.

[41] Sudhop, T; Reber, M; Tribble, D; et al. Changes in cholesterol absorption and cholesterol synthesis caused by ezetimibe and/or simvastatin in men. *J. Lipid. Res*, 2009, 50, 2117-2123.

[42] Katan, MB; Beynen, AC. Characteristics of human hypo- and hyperresponders to dietary cholesterol. *Amer. J. Epidemiol.*, 1987, 126, 387-399.

[43] Sudhop, T; Gottwald, BM; Bergman, K. Serum plant sterols as a potential risk factor for cardiovascular disease. *Metabolism*, 2002, 51, 1514-21.

[44] Sutherland, WHF; Williamns, MJA; Nye, NJ; Restieaux, NJ; deJong, SA; Walker, HL. Associations of plasma noncholesterol sterol levels with severity of coronary artery disease. *Nutr. Metab. Cardiovasc. Dis*, 1998, 8, 389-391.

[45] Weingartner, O; Lutjohann, D; Vanmierlo, T; et al. Markers of enhanced cholesterol absorption are a strong predictor for cardiovascular diseases in patiens without diabetes mellitus. *Chem. Phys. Lipids*, 2011, 164, 451-456.

[46] MacKay, DS; Johes, PJH. Plasma noncholesterol sterols: current uses potential and need for standardization. *Curr. Opin. Lipidol*, 2012, 23, 241-247.

[47] Assmann, G; Cullen, P; Ertey, J; et al. Elevation in plasma sitosterol concentration is associated with an increased risk for coronary events in the Procam Study. *Circulation*, 2003, 108, Suppl. 3300.

[48] MacKay, D; Jones, PJH. Evaluation of methods for the determination of cholesterol absorption and synthesis in humans. *Atherosclerosis*, 2011, 218, 253-262.

[49] Thompson, GR; O Neil, FO; Seed, M. Why some patients respond poorly to statins and how this might be remedied. *Europ. Heart J*, 2002, 23, 200-206.

[50] Folch, J; Lees, M; Sloane Stanley, GHS. A simple method for the isolation and purification of total lipides from animal tissues. *J. Biol. Chem*, 1957, 226, 497-509.

[51] Thompson, RH; Merola, GV. A simplified alternative to the AOAC official method for cholesterol in multi-component food. *AOAC Int*, 1993, 76, 1057-1068.

[52] Hušek, P. Chlormravenčany v analytické chemii jako činidla rychlé esterifikace. [Chloroformates in analytical chemistry as reagent of rapid esterification; available only in Czech]. *Unpublished PhD thesis*, 1996.

[53] Dayspring T. Noncholesterol sterols/stanols. The key to understanding cholesterol homeostasis: The lab report, Laboratory science bulletin, 2012. Available online at www.hdlainc.com/v 1i2/new-laboratory-tests.

[54] Hyanek, J; Pehal, F; Dubska, L; et al. Lathosterol and noncholesterolsterols in routine use for the differentiation and monitoring of dietary and drug induced treatment of hypercholesterolemias in children and adolescentes. *J. Nutr. Therap*, 2014, 3, 1-12.

In: Phytosterols
Editor: Deanna Garner
ISBN: 978-1-63483-477-3


*Chapter 7*

# PHYTOSTEROLS AND HEALTH BENEFITS: GC-MS DETERMINATION OF SURROGATE MARKERS FOR CHOLESTEROL SYNTHESIS/ABSORPTION IN PORTUGUESE YOUNG PEOPLE

***Rute Semedo[1], David Saraiva[1], Maria da Conceição Castilho[1], José Manuel Silva[2] and Fernando Ramos[1,3*]***

[1]Faculty of Pharmacy, University of Coimbra, Azinhaga de Santa Comba, Coimbra, Portugal

[2]Department of Internal Medicine, Coimbra University Hospitals, Faculty of Medicine, University of Coimbra, Coimbra, Portugal

[3]CNC-Center for Neuroscience and Cell Biology, Faculty of Pharmacy, University of Coimbra, Azinhaga de Santa Comba, Coimbra, Portugal

## ABSTRACT

Early development of cardiovascular diseases is clearly associated with high plasma or serum cholesterol concentrations. Cholesterol homeostasis is reflected by the balance between cholesterol absorption *vs.* cholesterol synthesis. The ratios to serum cholesterol of sterol precursors reflect synthesis rate of cholesterol, while those of phytosterols reflect cholesterol absorption. A study with 95 volunteer healthy students from Coimbra University was developed in order to determine the prevalence of surrogate markers for cholesterol synthesis and absorption. Blood was collected and centrifuged and the serum was submitted to alkaline saponification, followed by *n*-hexane extraction and subsequently evaporation to dryness under a nitrogen stream. The dry residue was derivatized and the cholesterol and non-cholesterol sterols were determined by Gas Chromatography-Mass Spectrometry. Obtained results showed two relevant facts. Firstly, total cholesterol concentrations in the study population were quite high (21% of analyzed students with >200 mg/dL), given the age group (19 to 24 years). Secondly, the

*Corresponding author: email. framos@ff.uc.pt.

cholesterol absorption markers (phytosterols) were found in concentrations higher than the markers of synthesis.

Obtained results can become an important contribution for developing better treatments to maintain healthy cholesterol concentrations, either for the elaboration of dietetic plans, namely for Mediterranean young people.

**Keywords:** cardiovascular diseases, cholesterol, campesterol, sitosterol, desmosterol, lathosterol

## INTRODUCTION

Cardiovascular diseases (CVD) are, nowadays, a major cause of mortality in Europe and in the USA [1]. Many factors contribute to the premature onset of CVD, but studies have shown that more than 90% risk of development can be attributed to high cholesterol concentrations, smoking and hypertension [2, 3]. Prevention of CVD is directly related to the reduction of plasma LDL-cholesterol levels. Knowing that the control of cholesterol levels may be crucial in the prevention and regression of CVD, it seems to be appropriate to conduct a study to better understand this finding, as well as to optimize and validate the methodologies currently employed in these kinds of studies [3-8].

The accumulated evidence in recent years suggest that the quantification of β-sitosterol, desmosterol, lathosterol, campesterol and other sterols in serum may be useful to identify subgroups of patients with high or low absorption and/or cholesterol synthesis. Cholesterol homeostasis is reflected by the ratio of cholesterol absorption *vs.* cholesterol synthesis. The ratios to serum cholesterol of cholesterol precursors (ex. desmosterol and lathosterol) reflect synthesis rate of cholesterol, while those of absorption of phytosterols (ex. campesterol and β-sitosterol) reflect cholesterol absorption efficiency [9-15]. Thus, cholesterol precursors and phytosterols are two key factors in understanding the cause of high levels of serum cholesterol and may be useful to reveal subgroups of patients with high or low absorption and/or synthesis of cholesterol.

The assessment of sterols and cholesterol in blood serum can predict the effectiveness of statin therapy in reducing cholesterol synthesis and may help to identify individuals with high cholesterol absorption and low synthesis [16, 17]. This will allow patients to benefit from combination therapy with an inhibitor of cholesterol absorption.

Most of the methods used so far in sterols analysis in serum are based on cholesterol determinations. Determination of serum cholesterol, its precursors and phytosterols are usually carried out by Gas Cromatography (GC), although High Performance Liquid Chromatography (HPLC) could also be utilized [7, 8,18-20]. However, for the referred compounds, GC has the leading advantage of superior resolution [21]. When coupled to mass spectrometry (MS), GC is the methodology that gives the most effective phytosterols and cholesterol precursor's resolution, identification and quantification [7, 19].

Previous studies have shown that serum cholesterol concentrations are relatively high in young people, decrease slightly in puberty, and start to increase again in young adults [22, 23]. Nevertheless, no studies characterizing the young population of Portugal in relation to cholesterol homeostasis (synthesis *vs.* absorption) are available. Thus, after the development of the analytical method of cholesterol and non-cholesterol sterol in human blood using GC-MS published in a previous paper [19], this study is the a contribution for the assessment of synthesis and absorption cholesterol markers in a young population, using Portuguese young people as a model.

# EXPERIMENTAL METHODS

## Study Population

This study was conducted according to the guidelines laid down in the Declaration of Helsinki and all procedures involving human subjects were approved by the Coimbra University Hospitals Ethical Committee. Written informed consent was obtained from all the 95 participants enrolled in the study (for more details about study population see Table 1).

To participate in the study, students must be in good health and not be taking any medication except oral contraceptives in the case of female participants. Diets of all participants are free of enriched functional food in plant sterols.

## Sterols Determination

Serum was obtained in the morning (9-11h) by blood collection from the flexure vein in the right arm of the 95 healthy adult students. After centrifugation of each individual blood sample at 2600±100 *g* for 10 min at room temperature, the obtained serum was transferred into vial and stored in the dark at -20ºC until analysis. Sterols determination was described previously [19]. Briefly, the serum aliquot was submitted to saponification by potassium hydroxide solution. The unsaponificable fraction was extracted with *n*-hexane that was evaporated to dryness through a gently nitrogen stream. The dry residue was derivatized and sterol determinations were made by gas chromatography-mass spectrometry. Samples were analyzed in triplicate.

**Table 1. Participants data according cholesterol groups classification**

| | **Normal** | | **Threshold** | | **High** | |
|---|---|---|---|---|---|---|
| | **female** | **male** | **female** | **Male** | **female** | **male** |
| Participants (n) | 65 | 10 | 12 | 3 | 4 | 1 |
| Total cholesterol (mg/dL)* | 168.0 ± 22.2 | | 216.7 ± 12.3 | | 273.1 ± 16.0 | |
| Age (years)* | 21.6 ± 1.4 | | 21.3 ± 2.7 | | 20.0 ± 1.1 | |
| Weight (kg)* | 60.1 ± 7.9 | | 54.7 ± 7.3 | | 59.4 ± 8.2 | |
| Height (m)* | 1.68 ± 0.09 | | 1.65 ± 0.08 | | 1.68 ± 0.09 | |
| BMI* | 21.3 ± 2.5 | | 20.8 ± 3.0 | | 21.2 ± 1.5 | |

*average ± standard deviation.

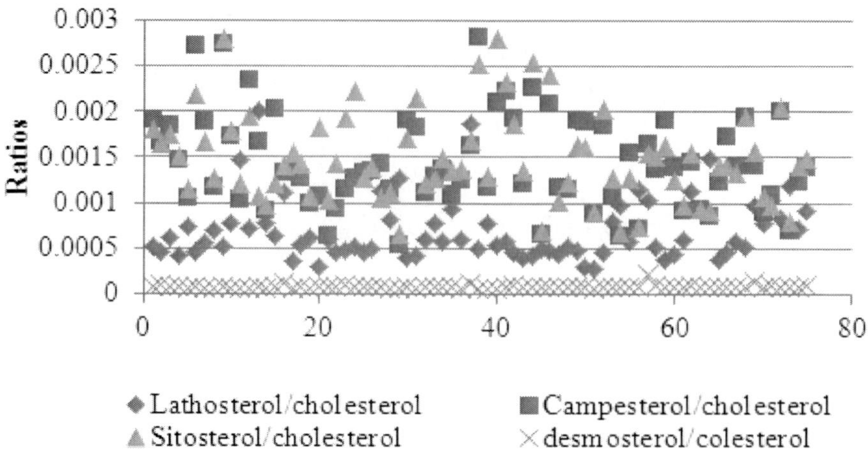


Figure 1. Sterol ratios of normal cholesterol group level.

## Results and Discussion

Obtained data were classified according to usual reference parameters of blood total cholesterol: normal values are below 200 mg/dL, the threshold values up to 239 mg/dL and higher values from 240 mg/dL [5, 6].

The population sample analyzed consisted of 95 Portuguese university young students, whose cholesterol levels should not exceed 200 mg/dL. However, it appears that although 75 of the values are below normal level, the majority of the cholesterol concentrations in blood are very close to the referred level (200 mg/dL), while 20 students present total cholesterol values higher the normal level, 5 of them with marked hypercholesterolemia values (>240 mg/dL). The total cholesterol levels of the student population do not seem to be related with gender nor with body mass index (BMI), despite the limited number of analyzed samples. Average data of total cholesterol, age, weight, height and BMI of the participants are shown in Table 1. Thus, for a better evaluation, a comparison of cholesterol synthesis and absorption markers adjusted for cholesterol and expressed as ratios was carried out among normal, threshold and high total cholesterol group levels. Nevertheless, one general observation could already be done: the ratio desmosterol/cholesterol does not provide adequate information, taking into account the methodology used in the analysis of data. Desmosterol presents very low values and the respective desmosterol/cholesterol ratios were very small (Figures 1, 2 and 3), confirming option of Lupattelli and co-workers [15] that only used lathosterol as cholesterol synthesis surrogate biomarker. On the other hand, it seems possible to conclude that, in general, the ratios indicate that the absorption of cholesterol (β-sitosterol and campesterol/cholesterol) are higher than that indicate the correspondent synthesis in the present study (lathosterol/cholesterol), as Miettinen and co-workers had already found in Finnish boys [24].

Although most young people present absorption ratios higher than those of synthesis, is possible to see in Figure 1 (normal group) that eight of them are different (cholesterol synthesis markers > absorption markers). That is, about 10% of young people already seem to have a high capacity to biosynthesis cholesterol, probably due to an individual genetic variation.

In the threshold group (Figure 2), according to the results presented, it appears clearly that the ratio campesterol/cholesterol and β-sitosterol/cholesterol concentration show(s) higher values. The values of lathosterol/cholesterol are inconsistent, showing that in one sample synthesis is greater and in other 3 individuals the ratios indicate that the absorption and synthesis ratios are very similar. In general the absorption prevails in this group.

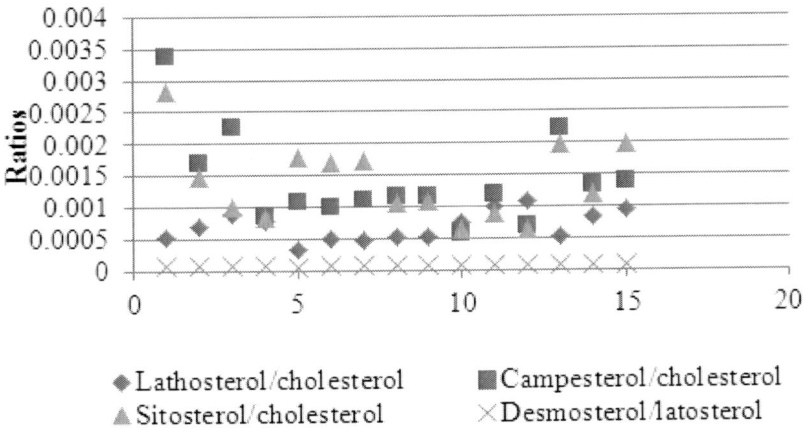


Figure 2. Sterol ratios of threshold cholesterol group level.

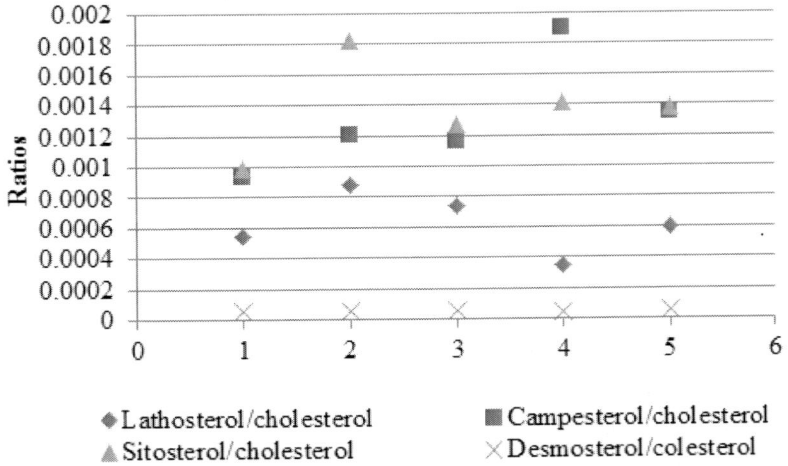


Figure 3. Sterol ratios of high cholesterol group level.

The results led to make a statement, an interesting speculation, if high group was exclude about the different groups on the synthesis and absorption markers, leading to an interesting speculation: contrary to what was supposed, if high group was exclude, the ratios variation were not constant, in that an increase in the ratio of a marker synthesis was not accompanied by a decrease in the ratio of absorption and vice versa. In general, the absorption markers were found in concentrations higher than the respective of synthesis markers, which suggest that the young people are, mainly, absorbers of cholesterol.

The results obtained in the high group (Figure 3), despite the small sample quantity, seem to be a clear indication that the referred students are mainly absorbers of cholesterol.

Preliminary conclusions about the prevalence of absorbers and/or synthesizers of cholesterol in each group of young people could be drafted, as above mentioned. This knowledge will certainly contribute to advances in the application of different hypercholesterolemia therapeutic, namely suggesting that for optimization of the therapy a case-to-case approach is necessary.

The higher serum cholesterol levels in young population are probably due to a high intake of dietary fat and cholesterol, but genetic factors could also be involved. However, it seems that the obtained data encourage further studies on a much larger and representative population taking account to establish an adequate program of food education in young people in order to reduce the incidence of high cholesterol blood levels as a means of preventing cardiovascular diseases.

In conclusion, this preliminary study with Portuguese university students can become an important contribution for developing better treatments to maintain healthy cholesterol concentrations, either for the prescription of therapeutic drugs or for the elaboration of dietetic plans, namely for young people living in a Mediterranean country.

## ACKNOWLEDGEMENTS

The authors thank all the volunteers who participated in the study. Research funds were provided by the Becel Institute (PT) and by the Portuguese Foundation for Science and Technology (Grant POCTI-SFA-8-177).

## REFERENCES

[1] Leif, R; Erhardt, A; Gotto, J. The evolution of European guidelines: Changing the management of cholesterol levels. *Atherosclerosis*, 2006, 185, 12-20.

[2] Nicholls, J; Wolski, K; Sipahi, I; et al. Rate of Progression of Coronary Atherosclerotic Plaque in Women. *J Am Coll Cardiol*, 2007, 49, 1546-1551.

[3] Callow, D. Cardiovascular disease: the global Picture. *Vascul Pharmacol*, 2006, 45, 302-307.

[4] Tonkin, M. Cholesterol and Coronary Heart Disease. *Asia Pacif Heart J*, 1996, 5, 105-110.

[5] NCEP (National Cholesterol Education Program): *Detection, Evaluation, and Treatment of High Blood Cholesterol in Adults (Adult Treatment Panel III): final Report, Circulation*, 2002, 106, 3143-3421.

[6] Perk, J; De Backer, G; Gohlke, H; et al. European guidelines on cardiovascular disease prevention in clinical practice (version 2012). *Eur Heart J*, 2012, 33, 1635-1701.

[7] Andrade, I; Santos, L; Ramos, F. Advances in analytical methods to study cholesterol metabolism: the determination of serum noncholesterol sterols. *Biomed Chromatogr*, 2013, 27, 1234-1242.

[8] Mackay, D; Jones, PJH. Evaluation of methods for the determination of cholesterol absorption and synthesis in humans. *Atherosclerosis*, 2011, 218, 253-262.

[9] Gylling, H; Hallikainen, M; Nissinen, MJ; et al. Very high plant stanol intake and serum plant stanols and non-cholesterol sterols. *Eur J Nutr*, 2010, 49, 111-117.

[10] Miettinen, TA; Tilvis, RS; Kesäniemi, YA. Serum plant sterols and cholesterol precursors reflect cholesterol synthesis and absorption in volunteers of a randomly selected male population. *Am J Epidemiol*, 1990, 131, 20-31.

[11] Nissinen, MJ; Gylling, H; Miettinen, TA. Responses of surrogate markers of cholesterol absorption and synthesis to changes in cholesterol metabolism during various amounts of fat and cholesterol feeding among healthy men. *Br J Nutr*, 2008, 99, 370-378.

[12] Weingärtner, O; Lütjohann, D; Böhm, M; et al. Relationship between cholesterol synthesis and intestinal absorption is associated with cardiovascular risk. *Atherosclerosis*, 2010, 210, 362-365.

[13] Miettinen, TA; Gylling, H; Nissinen, MJ. The role of serum non-cholesterol sterols as surrogate markers of absolute cholesterol synthesis and absorption. *Nutr Metab Carbiovasc Dis*, 2011, 21, 765-769.

[14] Mackay, DS; Jones, PJH. Plasma noncholesterol sterols: current uses, potential and need for standardization. *Curr Opin Lipidol*, 2012, 23, 241-247.

[15] Lupattelli, G; Pirro, M; Mannarino, MR; et al. Visceral fat positively correlates with cholesterol synthesis in dyslipidaemic patients. *Eur J Clin Invest*, 2012, 42, 164-170.

[16] Miettinen, T; Gylling, H. Ineffective decrease of serum cholesterol by simvastatin in a subgroup of hypercholesterolemic coronary patients. *Atherosclerosis*, 2002, 164, 147-152.

[17] de Jong, A; Plat, J; Lütjohann, D; et al. Effects of long-term plant sterol or stanol ester consumption on lipid and lipoprotein metabolism in subjects on statin treatment. *Br J Nutr*, 2008, 100, 937-941.

[18] Ramos, F; Saraiva, D. Plant Sterols and Plant Stanols in Milk Products Used as Functional Foods: Effects on Cardiovascular Risk Diseases Prevention. *In: Milk Consumption and Health (Lango E, Vogel F, ed.) Nova Science Publishers, Inc.*, 2009, 1-42.

[19] Saraiva, D; Semedo, R; Castilho, MC; et al. Selection of the derivatization reagent – the case of human blood cholesterol, its precursors and phytosterols GC-MS analyses. *J Chromatogr B*, 2011, 879, 3806-3811.

[20] Ito, N; Hakamata, H; Kusu, F. Simultaneous determination of beta-sitosterol, campesterol, stigmasterol, and brassicasterol in serum by high-performance liquid chromatography with electrochemical detection. *Anal Methods*, 2010, 2, 174-179.

[21] Field, F. Coupling chemical derivatization reactions with supercritical fluid extraction. *J Chromatogr A*, 1997, 785, 239-249.

[22] Akerblom, H; Viikari, J; Räsänen, L; et al. Cardiovascular risk in young Finns; results from the second follow-up study. *Ann Med*, 1989, 21, 223-225.

[23] Viikari, J; Rönnemaa, T; Seppänen, A; et al. Serum lipids and lipoproteins in children, adolescents and young adults in 1980-1986. *Ann Med*, 1991, 23, 53-59.

[24] Miettinen, T; Gylling, H; Viikari, J; et al. Synthesis and absorption of cholesterol in Finnish boys by serum non-cholesterol sterols. The cardiovascular risk in young Finns study. *Atherosclerosis*, 2008, 200, 177-183.

In: Phytosterols
Editor: Deanna Garner

ISBN: 978-1-63483-477-3


*Chapter 8*

# INFLUENCE OF FACTORS THAT MODULATE EFFICACY OF PHYTOSTEROLS

***Milagros Rocha[1,2,3,*], Celia Bañuls[1,2], Susana Rovira-Llopis[1,2], Carlos Morillas[1,2], Silvia Veses[1], Sandra López-Domenech[1], Noelia Díaz-Morales[1], Víctor Manuel Víctor[1,2,3,4] and Antonio Hernández-Mijares[1,2,5]***

[1]Service of Endocrinology. University Hospital Dr. Peset, Foundation for the Promotion of Health and Biomedical Research in the Valencian Region (FISABIO), Valencia, Spain
[2]Institute of Health Research INCLIVA, Valencia, Spain
[3]CIBER CB06/04/0071 research group. CIBER Hepatic and Digestive Diseases, Valencia, Spain
[4]Department of Physiology, Faculty of Medicine, University of Valencia, Valencia, Spain
[5]Department of Medicine, Faculty of Medicine, University of Valencia, Valencia, Spain

## ABSTRACT

**Background:** Phytosterols (PS), which are structurally similar to cholesterol, reduce intestinal cholesterol absorption and have other potentially beneficial effects in humans. Based on recent meta-analyses, a PS intake of 2 g/day lowers LDLc by an average of 8-10%, though this effect varies depending on intrinsic and extrinsic factors particular to the individual. We have assessed whether baseline cholesterol concentration, amount of cholesterol in the diet, Apolipoprotein (Apo) E genotype, and presence of metabolic syndrome (MetS) influence the effectiveness of PS. Furthermore, we have evaluated whether a decrease in carotenoids associated with consumption of a therapeutic dose of PS alters oxidative stress parameters.

**Material and methods:** This dietary intervention study was performed in moderate hypercholesterolemic and normocholesterolemic subjects. The study was of a parallel, randomized design with three arms, and was conducted over a period of 6 months. The first arm was performed with 52 volunteers from the general population who continued

* Corresponding author: Milagros Rocha, University Hospital Dr. Peset-FISABIO, Av. Gaspar Aguilar 90, 46017 Valencia, Spain. e-mail: milagros.rocha@uv.es, Tel: (34) 961622492. Fax: (34) 961622492.

their usual diet plus 2 g PS/day administered in 500 ml of low fat milk for 3 months. In the other two arms, 75 hypercholesterolemic individuals were administered a healthy diet for 3 months in order to standardize dietary habits. After this adjustment, subjects were divided into two intervention groups: one group followed a healthy diet with 2 g of PS/day for 3 months, and the other group followed the healthy diet without PS enrichment. Anthropometric parameters, lipid profile, Apo E genotype, oxidative stress parameters (susceptibility to oxidation in LDL particles, total antioxidant capacity and lipid peroxidation) and concentrations of serum carotenoids and PS (β-sitosterol and campesterol) were measured at the beginning and end of the study.

**Results:** Three-month dietary supplementation with PS produced a significant decrease in blood cholesterol (5.3% in total cholesterol and 6.5% in LDLc) in the general population without dietary restrictions. The reduction of LDLc was more pronounced in subjects with higher baseline concentrations of cholesterol. The beneficial effect of PS on the lipid profile of mildly hypercholesterolemic patients was not affected by the amount of saturated fat or cholesterol in their diet, nor was it mediated by the genotype of the Apo E. PS therapy appeared to be of little value to MetS patients, probably because of the reduced intestinal absorption of cholesterol. On the other hand, consumption of PS did not reduce serum total antioxidant capacity or increase oxidative stress.

**Conclusions:** Our results demonstrate that the efficacy of PS does not depend on the composition of macronutrients in the diet or the apolipoprotein E genotype; on the contrary, it is subject to baseline levels of LDL cholesterol. Moreover, PS therapy appears to be of little value to MetS patients.

## INTRODUCTION

Elevated concentrations of plasma low-density lipoprotein cholesterol (LDLc) are a major risk factor for the development of premature cardiovascular disease (CVD). According to data from the World Health Organization, CVDs are responsible for 17.1 million deaths each year (Lloyd-Jones et al., 2010). The first step towards reducing LDLc levels in individuals with mild to moderate hypercholesterolemia is to modify lifestyle and diet, in particular by reducing the intake of total and saturated fats (Expert Panel on Detection, Evaluation, and Treatment of High Blood Cholesterol in Adults, 2001). The effects of such a measure can vary, with a mean decrease in LDLc of between 5% and 10% being achieved (Hunninghake et al., 1993; Davidson et al., 1996). Among other recommendations, daily consumption of phytosterols (PS) - including plant sterols and plant stanols - has been shown to reduce the plasma concentration of LDLc by 10–15% (Katan et al., 2003; von Bergmann et al., 2005; Ras et al., 2013) without altering high-density lipoprotein cholesterol (HDLc) or triglyceride (TG) levels (Noakes et al., 2005; Seppo et al., 2007). The primary dietary sources of PS are fat-rich vegetables including vegetable oils and nuts - foods that are usually consumed in the Mediterranean region at maximum concentrations of 375 mg/day (Jimenez-Escrig et al., 2006) - although, to obtain therapeutic effects, patients are recommended a daily intake of PS in the range of 1–2 g/day (Expert Panel on Detection, Evaluation, and Treatment of High Blood Cholesterol in Adults, 2001), which should be accompanied by dietary measures (Cleghorn et al., 2003).

The mechanism by which dietary PS exert hypocholesterolemic effects is not fully understood. However, based on their physicochemical properties, it has been postulated that PS act at different levels. At the intestinal level, PS undermine the absorption of cholesterol from both the diet and enterohepatic circulation by competing with cholesterol molecules for uptake into mixed micelles (von Bergmann et al., 2005, Calpe-Berdiel et al., 2009), resulting in a reduction in the amount of cholesterol hydrolyzed and available for absorption (Awika and Rooney, 2004). Additionally, as cholesterol absorption decreases, synthesis increases, and PS intake has been found to increase endogenous cholesterol synthesis in humans (Jones et al., 2000). In the enterocyte, PS may inhibit cholesterol esterification at the level of acyl-coenzyme A cholesterol acyl transferase(ACAT) activity and increase the activity and expression of the type ABC carrier, accelerating cholesterol efflux from intestinal cells into the intestinal lumen.

Despite the abundance of clinical trial data demonstrating the cholesterol-lowering action of PS supplementation, a substantial variability in efficacy has been reported across individuals (Jones, 2015). Factors that might influence the cholesterol-lowering action of PS include their form, matrix and frequency of intake, and type of diet consumed. Most previous reports have suggested that the dispersion of PS in different food forms substantially affects the degree of LDLc reduction achieved (Katan et al., 2003). In relation to frequency, most studies have shown that PS are more effective when consumed in two to three portions per day (De Graaf et al., 2002; Mensink et al., 2002; Thomsen et al., 2004) than when a single dose is administered (Plat et al., 2000, Abumweis et al., 2006). Plat et al. (2000) found that a single dose of PS consumed at lunch had the same effect as that produced when a dose was provided at each meal time (three times per day), whereas a single dose of different preparations of PS administered in the morning was reported to lower LDLc levels (Abumweis et al., 2006). However, one of the most obvious factors at play here is the baseline cholesterol levels of individuals and how they may influence the capacity of PS to influence lipid profile.

Results from various studies are inconsistent, with baseline LDLc levels having been shown to modify the effect of PS in some (Mussner et al., 2002; Naumann et al., 2003), but not in others (Maki et al., 2001; Weststrate and Meijer, 1998). Abumweis et al. (2008) published a meta-analysis of fifty-nine randomised clinical trials. They concluded that baseline LDLc levels had affected the magnitude of the reduction in LDLc after PS consumption, which may explain the wide variation in responsiveness reported in previous studies. Demonty et al. (2009) obtained similar results and explored the effect of age on efficacy. The regression analysis performed in their meta-analysis, which did not detect a significant effect of age when baseline LDLc concentrations were included in the model, confirmed that the larger reduction in older subjects was due mainly to their higher baseline LDLc concentration. In another meta-analysis, Wu et al. (2009) also observed that the LDLc-lowering efficacy of PS varied according to baseline lipid levels. Finally, in a meta-analysis of five studies undertaken within the same laboratory to minimize sample analysis variation, consumption of plant stanol ester resulted in larger absolute decreases in total cholesterol and LDLc when patients had higher baseline serum levels, though the percentage of decrease in cholesterol was unaffected (Naumann et al., 2008).

Other potential modifier of the effect of PS supplementation on LDLc levels could be of a genetic profile (Jones, 2015). Apolipoprotein E (Apo E) is a major protein constituent of triglyceride-rich lipoprotein particles (chylomicron and very low-density lipoprotein cholesterol (VLDLc)) and their remnants, and also of HDLc. It serves as a ligand for the uptake of these lipoprotein particles to their receptors (Mahley, 1988). However, such functions are not uniformly effective, as Apo E is highly polymorphic, with three common alleles (e2, e3, e4) codifying its three main isoforms (E2, E3, E4). These proteins determine changes in Apo E plasma concentrations and differ in their affinity to their specific receptors.

While there are clear epidemiological data relating the Apo E genotype with CVD risk, evidence supporting the association with serum lipid variation is not nearly as convincing. There are data which indicate that the Apo E genotype is responsible for 7% of serum lipid concentration variations across individuals, whereas the results of the Framingham offspring study demonstrated a mere 1.0% variance and 2.1% variance in LDLc levels in men and women, respectively, in response to the Apo E locus (Schaefer et al., 1994; Ordovas, 2009). Despite this, observational population studies have demonstrated that Apo E2 phenotypic subjects have statistically lower total cholesterol and LDLc concentrations than those with Apo E3 and Apo E4 phenotypes, and that Apo E4 phenotypic subjects have statistically higher levels of LDLc than their Apo E3 counterparts (Schaefer et al., 1994; Howard et al., 1998; Plat and Mensink, 2002; Nissinen et al., 2008). In reference to the effect of PS, some studies have indicated that PS reduce cholesterol more effectively in individuals with the Apo E4 genotype than those with the Apo E3 genotype, thus pointing to an Apo E genotypic effect (Weststrate and Meijer, 1998; Vanhanen et al., 1993; Miettinen and Vanhanen, 1994). On the contrary, Sanchez-Muniz et al. (2009) concluded that PS supplementation is of little value in Apo E4 carriers. However, most studies have failed to find significant variation in the effects of PS depending on genotype.

An extrinsic factor that could also determine the efficacy of PS includes background diet. The level of dietary cholesterol intake in conjunction with PS treatment may alter the inhibitory effect of PS on cholesterol absorption, since higher intake may improve the efficacy of PS. In fact, administration of PS to gerbils fed different amounts of cholesterol resulted in an interaction between dietary cholesterol and PS activity (Hayes et al., 2002). In humans, the cholesterol-lowering efficacy of PS has been independently studied with respect to both low and high levels of cholesterol intake, with similar reductions in LDLc and total cholesterol being obtained. Reductions of 4% to 9% in total cholesterol and 8% to 16% in LDLc were observed when varying doses of PS were administered as part of a low-cholesterol diet (Jones et al., 2000; Maki et al., 2001; Devaraj et al., 2004). Similarly, reductions of up to 9% and 15% in total cholesterol and LDLc, respectively, were reported as a result of a high-cholesterol diet including PS (Vanstone et al., 2002; Quilez et al., 2003). However, only a few studies have directly compared high- and low-cholesterol diets in terms of their effect on PS activity. While some investigators have reported that low levels of dietary cholesterol intake attenuate the effectiveness of PS (Denke, 1995; Mussner et al., 2002), others have shown that PS lower cholesterol concentration, even in subjects with a highly restricted cholesterol intake (Hallikainen and Uusitupa, 1999, Kassis et al., 2008).

The relationship between baseline markers of cholesterol absorption in lipid response and antiresorptive therapies (stanols/sterols) is well established, with stronger responses being associated with markers of cholesterol absorption (Gylling and Miettinen, 2002; Thuluva et al., 2005; Fuentes et al., 2008). There is firm evidence that the efficiency of cholesterol absorption is lower and that of cholesterol synthesis higher in insulin-resistant states such as metabolic syndrome (MetS) (Pihlajamäki et al., 2004; Ooi et al., 2009; Simonen et al., 2000). MetS englobes a cluster of metabolic disorders including dyslipidemia, abdominal obesity, hypertension and impaired fasting glucose. Individuals suffering these disorders commonly manifest elevated cholesterol synthesis and reduced cholesterol absorption (Simonen et al., 2000, Gylling et al., 2007). It has been speculated that the lower absorption of cholesterol observed in MetS patients interferes with the mechanism of action of PS, thus reducing their efficacy as hypocholesterolemic agents. However, there is a lack of relevant data, and the few studies carried out to assess the impact of PS supplementation on MetS subjects have obtained contradictory results. In one study, PS did not influence plasma lipid or lipoprotein metabolism in patients with MetS (Ooi et al., 2007), while, in another, Plat et al. (2009) reported a reduction of non-HDLc and TG in patients with overt hypercholesterolaemia and hypertriglyceridemia, despite the fact that apolipoprotein B (Apo B) remained unaltered.

An important issue regarding the efficacy and safety of PS consumption is the concomitant decrease in plasma levels of fat-soluble vitamins, particularly tocopherol and carotenoids. Since carotenoids and tocopherol are primarily transported within lipoproteins - mainly VLDL and LDL - alterations of lipoprotein fractions and total cholesterol may explain decreases in the mean concentration of antioxidants. According to Katan et al.(2003)'s meta-analysis of eighteen trials, PS lower α-carotene, β-carotene and lycopene serum concentrations.

One way to counterbalance this reduced absorption is to increase dietary intake of carotenoids. Some reports have endorsed adding carotenoids to products enriched with PS (Quílez et al., 2003; Brufau et al., 2004), whereas others have obtained excellent results after supplementing an enriched diet of fruit and vegetables with PS (Clifton et al., 2004a, Noakes et al., 2002, Colgan et al., 2004; Ntanios and Duchateau, 2002). On the contrary, other authors have failed to observe any significant reduction in carotenoids after administering PS in a food matrix naturally enriched with vitamins, namely orange juice (Devaraj et al., 2006). It is possible that this decrease undermines antioxidative defences and, therefore, increases oxidative stress. Oxidation of LDL particles has been implicated at both early and late stages of the pathogenesis of atherosclerosis, during which plaque rupture leads to further clinical events. The attack of reactive oxygen species against biomembranes or lipoproteins leads to oxidative destruction of polyunsaturated fatty acids by a process called lipid peroxidation, of which malondialdehyde (MDA) is a product and, therefore, an indicator of oxidative stress in cells and tissues. In contrast, due to the decrease in serum concentration of LDLc that is also produced by PS consumption (caused by a drop in the number of circulating LDL particles susceptible to oxidation), atherosclerotic lesions are less likely to form. Thus, whether a PS-mediated decrease in fat-soluble antioxidants leads to a global impairment of antioxidative defences or an enhancement of oxidative stress is still unclear.

In light of the abovementioned, the objectives of our research were to analyze, over a long-term period, the efficacy of daily consumption of 2 g of PS (a) in the general population according to baseline serum cholesterol concentration and in hypercholesterolemic patients according to (b) Apo E genotype, (c) background diet, and (d) MetS as a pathological model of reduced cholesterol absorption. In addition, we have also evaluated whether possible side effects on carotenoids impair oxidative stress parameters.

# Material and Methods

## Study Population

This dietary intervention study was performed in the general population and in moderate hypercholesterolemic subjects. Both groups were recruited among clinical and laboratory staff at the University of Valencia and Outpatient's Department of the Endocrinology Service of the University Hospital Dr. Peset. Subjects between the ages of 18 and 76 years (inclusive) were eligible for inclusion in the study.

Further inclusion criteria for patients with moderate hypercholesterolemia were a serum LDLc concentration of between 160-190 mg/dl in patients with less than 2 cardiovascular risk factors, 130-160 mg/dl in patients that presented 2 or more cardiovascular risk factors, and a TG concentration of <400 mg/dl in any case. Cardiovascular risk factors were defined as: age (≥ 45 years in men and ≥55 years in women), a smoking habit, hypertension (≥140/90 mmHg), diabetes mellitus, HDLc concentration of <40mg/dl, and a family history of CVD. Exclusion criteria were pregnancy or lactation, change of oral contraceptives, severe disease, a history of chronic inflammatory disease, hypersensitivity to milk proteins, and lipid-lowering medication.

The study was conducted according to the guidelines laid down in the Declaration of Helsinki, and all procedures involving human subjects were approved by the Ethics Committee of the Hospital. Written informed consent was obtained from all patients. On the basis of these criteria, 52 volunteers (of which 29 were hypercholesterolemic) were included in the general population group and were assigned a free diet plus PS. On the other hand, 75 patients with moderate hypercholesterolemia were randomly assigned a healthy diet (n = 34), or healthy diet plus PS (n = 41).

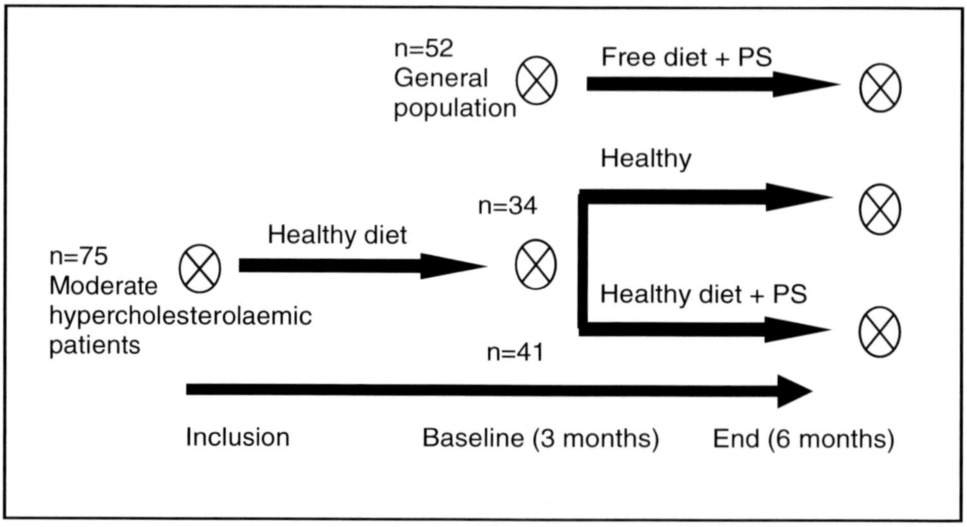


Blood samples were extracted at inclusion, at baseline and at the end of the experimental period.

Figure 1. Interventional clinical trial design to evaluate the efficacy of 2g/day of phytosterols (PS).

MetS is defined, according to the criteria of the American Heart Association/National Heart Lung and Blood Institute Scientific Statement, as the presence of at least three of the following risk factors (Grundy SM et al. 2005): (i)Waist circumference > 102 cm in men and > 88 cm in women ; (ii) TG ≥ 150 mg/dl; (iii) HDLc < 40 mg/dl in men and < 50 mg/dl in women; (iv) blood pressure (BP) ≥130/≥85 mmHg; (v) fasting glucose ≥100 mg/dl.

## Study Design

The study was a parallel, randomized design with three arms, and was conducted over a period of 6 months. The first arm was performed with 52 volunteers from the general population who continued their usual diet plus 2 g PS/day administered in 500 ml of low-fat milk for 3 months. In the other two arms, 75 moderate hypercholesterolemic individuals were administered a healthy diet for another 3 months in order to standardize dietary habits. After this adjustment, subjects were divided into two intervention groups: one group followed a healthy diet with 2 g of PS/day for 3 months, and the other group followed the healthy diet without PS enrichment (Figure 1).

Patients following a controlled diet were monitored throughout the trial by an experienced dietician and received detailed written and oral instructions about their diet, including the precise amounts and the quality of the food products to be consumed. The recommended diet included total cholesterol (<200 mg/day), saturated fat (<7% of daily total energy), simple sugar low contents (<10%), an enhanced consumption of monounsaturated fatty acids from olive oil (up to 20% of MUFA with respect to total energy), ω-3 fatty acids from fish, 2-3 pieces of fruit and unlimited vegetables. Insofar as daily energy intake, 8360 KJ (2000 Kcal) were proposed for men and 7106 KJ (1700 Kcal) for women, of which 18-19% was proteins, 52-53% was carbohydrates and 29-30% was fats, and which included 20-30 g of dietary fibre.

Adherence to the diet was monitored by means of 3-day food records (compiled on weekdays) and 24-hour diet recall at baseline, after the adaptation period and at the end of the study (carried out during appointments with the dietician). Food intake was converted into energy and nutrients with the help of the Spanish Food Composition Table (Mataix et al., 2003). The composition database was created with AYS44 Diet Analysis software obtained from ASDE, SA (Valencia, Spain). Patients were encouraged to maintain their normal pattern of activity.

The PS-enriched milk was produced by Unilever (Spain) and packed in white containers. In addition to 0.4 g of vegetal sterols, every 100 ml of milk provided 3.2 g protein, 4.7 g carbohydrates, 1.8 g fat (0.25 g saturated, 0.50 g monounsaturated, 1.05 g polyunsaturated) and 48 kcal of energy. The PS consisted of vegetable oil-based sterols esterified with sunflower oil fatty acids. The PS content of enriched milk was 70.7% β-sitosterol, 15.0% campesterol, 9.6% β-sitostanol, 2.6% brassicasterol, 1.0% campestanol and 0.9% stigmasterol. The diet group received commercially available low-fat milk with a similar macronutrient composition and energy intake to that consumed by the PS group but without PS. Patients were recommended to consume the milk twice per day with meals.

Compliance was evaluated by interviewing the patients and counting the unopened and unconsumed product packages returned to the clinic, and was recorded as the percentage of scheduled servings consumed. Non-compliance was defined as the consumption of <80 % of the scheduled servings during the study period.

## Blood Sampling

Venous blood samples were collected from moderate hypercholesterolemic patients after 12h overnight fasting at inclusion, after 3 months of healthy diet (Baseline), and at the end of the experimental period (End). In the case of free diet plus PS, blood samples were collected at baseline and after three months of PS consumption (Figure 1). To separate serum from blood cells, samples were immediately centrifuged at 2,000 g for 15 min at 4°C. Freshly separated serum was employed for determination of lipid profile, and the remaining aliquots of serum were stored at -80°C until they were assessed for further measurements.

## Lipid Analysis

Total cholesterol and TG were measured by means of enzymatic assays (Bucolo and David, 1973; Allain et al., 1974) and HDLc concentrations were recorded using a direct method (Sugiuchi et al., 1995) with a Beckman LX-20 autoanalyzer (Beckman Coulter, La Brea, CA, USA). The intraserial variation coefficient was <3.5% for all the determinations. When TG values were below 300 mg/dl, LDLc concentration was calculated using the Friedewald method (Friedewald et al., 1972). Non-HDLc concentration was obtained by calculating the difference between total and HDL cholesterol. Apo A and Apo B were determined by immunonephelometry (Dade Behring BNII, Marburg, Germany), for which the intra-assay variation coefficient was <5·5%. LDL size was evaluated after electrophoresis in polyacrylamide gradient-gel electrophoresis (2-16%), using the method described by Nichols et al. (1986).

## Apo E Genotype

To evaluate the effect of Apo E on their response to treatment, subjects were categorized into 3 groups -E2: E2/E2, E2/E3 and E2/E4; E3: E3/E3 and E4: E3/E4 and E4/E4-.

DNA was extracted for genotyping according to the method of Miller et al. (1988). Apo E was identified after PCR amplification of a 244 bp fragment using forward 5‘ -ACAGAATTCGCCCCGGCCTGGTACAC-3‘ and reverse 5‘-TAAGCTTGGCACG GCTGTCCAAGGA-3‘ as oligonucleotides. PCR reaction was performed in a 25 μl total volume containing 0.3 mM of each primer, 0.2 mM of each dNTP, 1x Taq buffer, 2 mM $MgCl_2$, 100 ng of genomic DNA and 2.5 U Taq polymerase (Netzyme) using a Master-Cycler thermocycler (Eppendorf Scientific, Inc., Westbury, NY, USA). Each reaction mixture was heated at 94°C for 30 s, 64°C for 30 s, and 72°C for 35 s. Forty cycles were performed. A final extension was carried out at 72°C for 5 min. The 8 μl PCR products were subjected to restriction enzyme analysis by digestion for 4 hours at 37°C with 0.4 U of the restriction endonuclease Hha I. The fragment products were then separated by polyacrylamide gel electrophoresis (8%). After electrophoresis, the gel was treated with silver staining and DNA fragments were visualized.

## PS Analysis

Serum campesterol and β-sitosterol concentration were measured by gas-liquid chromatography, as described by Wolthers, with minor modifications (Wolthers et al., 1991). In short, 2 μg of internal standard 5β-cholestan-3α-ol (Sigma-Aldrich, St Louis, MO, USA) was added to 0.1 ml serum samples that were then saponified with 1 ml of methanolic potassium hydroxide (0.71M) (Scharlau, Barcelona, Spain). Non-saponified materials were then extracted twice with hexane (Sigma-Aldrich, St Louis, MO, USA) and distilled water. The extracts were derivatized and injected into a gas-liquid chromatograph equipped with a flame ionization detector (Perkin Elmer Autosystem XL, Norwalk, USA) and a 30-m column (Supelco equity, Bellefonte, MO, USA). Campesterol and β-sitosterol serum concentrations were determined in triplicate by recording their peaks and expressing them in relation to the 5β-cholestan-3α-ol internal standard.

## Carotenoid Analysis

Serum concentrations of hydrocarbon carotenoids (β-carotene and lycopene) and oxygenated carotenoids (lutein and β-cryptoxanthin) were determined at baseline and the end of the trial by employing a HPLC method described by Olmedilla et al. (1990), with slight modifications. In brief, 400 μL of serum samples were mixed with an equivalent quantity of ethanol and an internal standard solution (0.6 μg/mL) of β-apo-8-carotenal. Fat-soluble compounds were extracted twice with n-hexane (1 ml each time), stabilized with 0.01 % butylated hydroxytoluene and the extract was allowed to completely evaporate under nitrogen. The samples were dissolved once more in 150 μL of tetrahydrofurane stabilized with BHT 0.01% for HPLC-reversed phase analysis (Waters, Eschborn, Germany). An isocratic method with dichloromethane:acetonitrile:methanol (20:70:10; v:v:v) at a flow of 1.3 ml/min was employed as a mobile phase. Detection was performed at 450 nm with AUFS 0.005.

## Oxidative Stress Parameters

LDLc oxidation was measured by monitoring the formation of conjugated dienes after in vitro $Cu^{2+}$-catalyzed oxidation. LDL cholesterol was isolated using a dextran sulfate LDL-adsorption column (Liposorber® LA-15 System Facts). Salts and EDTA were removed by passing the LDL samples through a polyacrylamide column (Bio-Gel P6). Samples were diluted to 100 μg LDL Apo B/ml with a phosphate-buffered saline (PH 7.4) and incubated at 37 °C with 5μM $Cu^{2+}$/l for 5h. Conjugated diene formation was measured by continuous spectrophotometric monitoring at 234 nm. Lag time was estimated as the incubation time corresponding with the intersection of 2 lines representing the changes in optical density; one through the initial, slowly rising curve, which represented the consumption of the endogenous antioxidants, and the other, rapidly rising curve, which represented the propagation phase of the LDL oxidation (Esterbauer et al., 1989).

Total antioxidant status was measured spectrophotometrically at 600 nm using the Trolox Equivalent Antioxidant Capacity (TEAC) assay (Randox, San Francisco, CA), a method developed by Miller et al. (1993), following the manufacturer's instructions. The intra- and inter-assay coefficients of variation were 1.7 and 3.5%, respectively.

Free radical formation through lipid peroxidation was determined using malondialdehyde-thiobarbituric acid reactive substances (MDA-TBARS), as described by Yagui (1998). In short, after separating proteins from MDA, a fluorescent complex (MDA-TBA) was formed and the fluorescence between the excitation and emission wavelengths (544-590 nm) was measured in a Fluoroskan Ascent FL (Thermo Electron Corporation).

## Statistical Analysis

This study was designed to have a power of 80%, in order to be able to detect differences among the groups in relation to the primary efficacy criterion (LDLc variation) equal to or greater than 15 mg/dl, assuming a common standard deviation of 23 mg/dl.

Data are expressed as mean and SD (for tables) or SEM (for figures) for parametric data. Statistical analysis was carried out with the statistics program SPSS version 17.0. The baseline characteristics of the groups were analyzed by an unpaired T test or Mann Whitney U test when two groups were compared or by one-way analysis of variance (ANOVA) or a Kruskal-Wallis test for parametric or non-parametric data, respectively. Between-group and within-group differences were analyzed using a 2-factor repeated-measures ANOVA followed by a paired Student's t test or one-way ANOVA for parametric data when no interaction was found. For non-parametric variables, data were normalized using a log transformation. $p < 0.05$ was considered significant.

**Table 1. Anthropometric and lipid parameters in the general population at baseline and after addition of phytosterols (PS) to their usual diet for three months**

| | Baseline | PS |
|---|---|---|
| Men/Women (n) | 22/30 | --- |
| Age (Yeats) | 40.6 ± 11.7 | --- |
| BMI (Kg/m$^2$) | 25.6 ± 3.5 | 25.3 ± 5.5 |
| Systolic BP(mm Hg) | 121 ± 17 | 119 ± 16 |
| Diastolic BP(mm Hg) | 78 ± 11 | 78 ± 10 |
| TC (mg/dl) | 209.8 ± 31.6 | 198.6 ± 38.4* |
| LDLc (mg/dl) | 138.5 ± 26.5 | 129.5 ± 32.2* |
| HDLc (mg/dl) | 52.5 ± 16.6 | 50.2 ± 17.9 |
| TG (mg/dl) | 85.6 ± 42.8 | 91.3 ± 85.7 |
| Apo B/Apo A ratio | 0.63 ± 0.19 | 0.58 ± 0.18* |
| LDL diameter (nm) | 26.1 ± 0.8 | 26.2 ± 0.8 |

Data are expressed as mean ± SD. Values of serum TG concentrations were normalized using a log transformation. * $p < 0.05$ when compared using a paired Student's t-test.

Abbreviations: BMI: body mass index; BP: blood pressure; TC: total cholesterol; LDLc: LDL cholesterol; HDLc: HDL cholesterol; TG: triglycerides; Apo: Apolipoprotein.

## RESULTS

To assess whether baseline serum cholesterol concentration influences the effectiveness of PS, we evaluated a total of 52 volunteers from the general population –22 men and 30 women- with a mean age of 41 ± 12 years. No significant differences were observed between anthropometric parameters at the end of the study (Table 1). However, three-month dietary supplementation with 2g of PS/day produced a significant reduction in total cholesterol (5.3%), LDLc (6.5%) and Apo B/Apo A ratio (7.9%), which was not associated with changes in LDL particle size (Table 1). Although the diameter of LDL particles and TG concentrations were unaltered after administration of PS, a negative correlation between these parameters was observed ($r = -0.367$; $p = 0.007$).

To assess in more depth alterations in LDLc concentration, they were analyzed according to LDLc tertiles at baseline (tertile 1: LDLc concentration less than 124 mg/dl; tertile 2: LDLc concentration between 124 and 154 mg/dl; tertile 3: LDLc concentration greater than 154 mg/dl), which showed PS to have been more effective in subjects with higher LDLc levels (Figure 2).

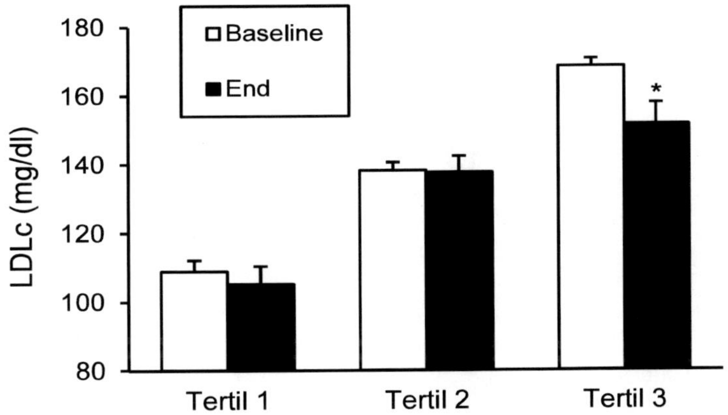


Data are represented as mean + SEM of 17-18 subjects per group. * $p < 0.05$ when compared using a paired Student's t-test (tertile 1: LDLc < 124 mg/dl; tertile 2: LDLc: 124-154 mg/dl; tertile 3: LDLc > 154 mg/dl).

Figure 2. LDLc concentration in the general population before and after addition of phytosterols (PS) to their usual diet for three months, according to tertiles of LDLc at baseline.

In light of these results, we evaluated the efficacy of a PS in a mild hypercholesterolemic population that followed a healthy diet throughout the experimental period. A total of forty subjects in the age range of 24-69 completed this study. Anthropometric parameters of the subjects before and after addition of PS to their diet are presented in Table 2. Since there were no significant differences between the diet and diet + PS groups at inclusion in age, weight, BMI, or systolic/diastolic BP, values are represented as the mean of the forty subjects. Men and women were distributed similarly between the two groups: 30% of the diet group and 25% of the diet + PS group were men (data not shown). During the first three months of the trial, subjects followed dietary measures to homogenize their plasma parameters prior to introducing PS into their diet. Diet significantly reduced weight and BMI in the entire population. No significant changes were observed between the anthropometric parameters of subjects receiving and those not receiving beverages enriched with PS (Table 2).

**Table 2. Anthropometric parameters and lipoprotein profile of hypercholesterolemic subjects before and after addition of phytosterols to their diet**

| | Inclusion | 3-month diet (Baseline) | Post-randomized | End |
|---|---|---|---|---|
| Age (years) | 50.0 ± 10.2 | ------------ | A: ------------ | A: ------------ |
| | | | B: ------------ | B: ------------ |
| BMI (Kg/m²) | 28.3 ± 4.7 | 27.7 ± 4.5* | A: 27.8 ± 4.6 | A: 28.0 ± 4.8 |
| | | | B: 27.6 ± 4.9 | B: 27.6 ± 5.2 |
| BP systolic (mmHg) | 131 ± 20 | 126 ± 15 | A: 129 ± 16 | A: 131 ± 15 |
| | | | B: 130 ± 13 | B: 130 ± 19 |
| BP diastolic (mmHg) | 82 ± 11 | 81 ± 14 | A: 86 ± 10 | A: 85 ± 10 |
| | | | B: 85 ± 10 | B: 86 ± 14 |
| TC (mg/dl) | 247 ± 24 | 237 ± 27* | A: 238 ± 30 | A: 232 ± 30 |
| | | | B: 236 ± 24 | B: 221 ± 29* |
| LDLc (mg/dl) | 171 ± 22 | 163 ± 22* | A: 164 ± 26 | A: 161 ± 28 |
| | | | B: 162 ± 19 | B: 146 ± 18* |
| HDLc (mg/dl) | 55 ± 12 | 54 ± 13 | A: 51 ± 12 | A: 50 ± 11 |
| | | | B: 58 ± 14 | B: 56 ± 15 |
| TG (mg/dl) | 107 ± 58 | 99 ± 52 | A: 112 ± 57 | A: 116 ± 65 |
| | | | B: 86 ± 44 | B: 95 ±69 |
| Apo B (mg/dl) | 117 ± 16 | 114 ± 16 | A: 120 ± 17 | A: 117 ± 16 |
| | | | B: 108 ± 11 | B: 105 ± 14 |

Data are expressed as means ± SD. Values of serum TG concentration were normalized using a log transformation. After the first three months of dietary therapy, forty subjects were randomized into two groups (20 subjects each group): one followed dietary therapy (A) and the diet of the other was enriched with 2g/d of plant sterols (B). * $p < 0.05$, when compared using a paired Student's t-test during the first three months of the study, or 2-factor repeated-measures ANOVA followed by paired Student t-test after addition of phytosterols.

Abbreviations: BMI: body mass index; BP: blood pressure; TC: total cholesterol; LDLc: LDL cholesterol; HDLc: HDL cholesterol; TG: triglycerides; Apo: Apolipoprotein.

During the first three months of the standard healthy diet, subjects showed a significant reduction of 4-5% in total cholesterol, LDLc and non-HDLc. Enrichment with PS induced an additional reduction of 6.4%, 9.9% and 7.1%, respectively, which represented a total drop from onset of the study of 10.5%, 14.6% and 14.1%, respectively. These lipid parameters were maintained in subjects undergoing dietary therapy only, therefore marking significant differences with the other group (Figure 3A, Table 2). Apo B/Apo A ratio, considered an even better marker of cardiovascular heart disease (CHD) than lipid concentration per se (Ridker et al., 2005), was reduced in the subjects receiving low-fat milk enriched with PS (Figure 3B). No changes were detected in HDLc, TG (Table 2) or LDL size (Figure 4) in any of the groups.

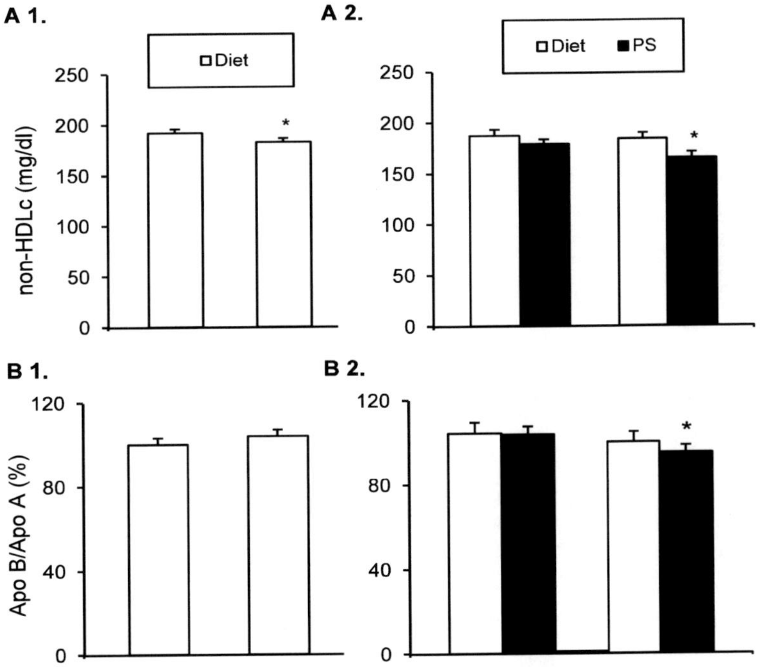


Panel A1 shows non-HDLc during the first 3 months, which corresponded with the diet standardization period.

Panel A2 shows HDL non-cholesterol during the following 3 months, which corresponded with the addition of phytosterols to the diet.

Panel B1 shows Apolipoprotein B/Apolipoprotein A ratio during the first 3 months, which corresponded with the diet standardization period.

Panel B2 shows the Apolipoprotein B/Apolipoprotein A ratio during the following 3 months, which corresponded with the addition of phytosterols to the diet.

Bars represent means ± SEM of 40 subjects in the first phase and 20 subjects in the second phase after addition of phytosterols to the healthy diet. * $p < 0.05$ when compared using a paired Student t-test in first three months of the study or 2-factor repeated-measures ANOVA followed by paired Student t-test after addition of phytosterols.

Abbreviations: Non-HDLc: non-HDL cholesterol; Apo: Apolipoprotein ; PS: phytosterols.

Figure 3. Cardiovascular risk parameters in hypercholesterolemic subjects before and after addition of phytosterols (PS) to their diet.

The whole hypercholesterolemic population was analysed in order to evaluate the efficacy of PS according to Apo E genotype. The baseline characteristics of participants according to their Apo E alleles are shown in Table 3. No differences in anthropometrical characteristics or lipoprotein profile were found in relation to any of the parameters studied. At the end of the three month intervention period, as we have previously explained, patients on the healthy diet supplemented with PS exhibited a significant reduction in total cholesterol, LDLc, non-HDLc and Apo B/Apo A ratio with respect to baseline conditions. Conversely, these lipid parameters were unaltered in patients undergoing dietary therapy only (Table 4). When patients were analyzed according to their Apo E alleles, we found that the effect of PS on lipid profile was independent of Apo E phenotype, with similar significant reductions being observed in total cholesterol, LDLc, non-HDLc and Apo B/Apo A ratio in patients with

E3 or E4 phenotypes (Table 4). Data related to the genotype E2 are not represented due to the negligible number of this type of patient in our study population (2 subjects in the group that followed a healthy diet and 4 subjects in the group receiving combined therapy).

The efficacy of PS, independently of Apo E polymorphism, is also presented in Figure 5. No differences were observed between patients with Apo E3 phenotype and those with Apo E4 phenotype in the percentage of decrease in total cholesterol, LDLc, non-HDLc or Apo B/Apo A after addition of PS to the standard healthy diet.

**Table 3. Baseline characteristics of hypercholesterolemic patients according to Apo E phenotype**

| | E2 | E3 | E4 | All |
|---|---|---|---|---|
| Subjects (n) | 6 | 47 | 22 | 75 |
| Male/Female (n) | 2/4 | 13/34 | 7/15 | 22/53 |
| Age (years) | 49.0 ± 12.3 | 50.1 ± 11.3 | 49.6 ± 13.7 | 49.9 ± 12.0 |
| BMI (Kg/m$^2$) | 27.9 ± 4.6 | 28.5 ± 4.4 | 27.7 ± 6.2 | 28.2 ± 5.0 |
| Systolic BP (mmHg) | 127 ± 12 | 129 ± 15 | 131± 18 | 130± 15 |
| Diastolic BP (mmHg) | 80 ± 15 | 82 ± 10 | 80 ± 15 | 81 ± 13 |
| TC (mg/dl) | 238 ± 26 | 237 ± 31 | 237 ± 28 | 237 ± 29 |
| LDLc (mg/dl) | 166 ± 20 | 164 ± 26 | 166 ± 20 | 165 ± 24 |
| HDLc (mg/dl) | 49 ± 9 | 50 ± 14 | 49 ± 13 | 50 ± 14 |
| Non-HDLc (mg/dl) | 188 ± 24 | 187 ± 29 | 188 ± 26 | 187 ± 27 |
| TG (mg/dl) | 114 ± 54 | 112 ± 62 | 108 ± 67 | 113 ± 62 |
| Apo B/Apo A ratio | 0.713 ± 0.096 | 0.728 ± 0.166 | 0.827 ± 0.191 | 0.757 ± 0.174 |

Data are expressed as mean ± SD. Values of serum TG concentrations were normalized using a log transformation. No differences among groups were revealed by one-way ANOVA.

Abbreviations: BMI: body mass index; BP: blood pressure; TC: total cholesterol; LDLc: LDL cholesterol; HDLc: HDL cholesterol; TG: triglycerides; Apo: Apolipoprotein.

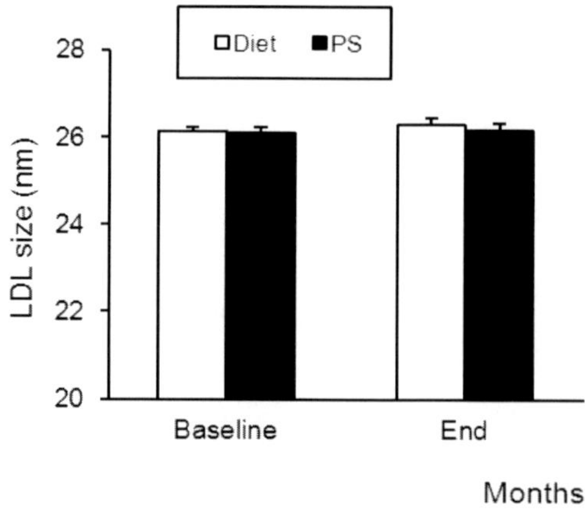


Data are represented as mean + SEM of 34 and 41 subjects in diet and phytosterols groups, respectively. No differences between groups were revealed by a paired Student's t-test.

Figure 4. LDL size in hypercholesterolemic subjects after the addition of phytosterols (PS) to their diet.

**Table 4. Serum lipoprotein profile of hypercholesterolemic patients after addition of phytosterols (PS) to their diet according to Apo E phenotype**

| | Diet | | | PS | | |
|---|---|---|---|---|---|---|
| | E3 | E4 | All | E3 | E4 | All |
| N | 23 | 9 | 34 | 24 | 13 | 41 |
| TC (mg/dl) | | | | | | |
| Baseline | 240 ± 36 | 246 ± 26 | 241 ± 32 | 234 ± 25 | 231 ± 29 | 234 ± 26 |
| End | 233 ± 32 | 244 ± 24 | 236 ± 30 | 225 ± 26* | 213 ± 27* | 221 ± 27** |
| LDLc (mg/dl) | | | | | | |
| Baseline | 164 ± 31 | 176 ± 16 | 162 ± 39 | 165 ± 21 | 160 ± 20 | 164 ± 21 |
| End | 159 ± 30 | 177 ± 19 | 164 ± 26 | 151 ± 20** | 145 ± 22.3* | 149 ± 21*** |
| HDLc (mg/dl) | | | | | | |
| Baseline | 50 ± 12 | 52 ± 14 | 50 ± 12 | 51 ± 16 | 47 ± 13 | 50 ± 15 |
| End | 50 ± 11 | 47 ± 7 | 49 ± 10 | 54 ± 16 | 46 ± 13 | 51 ± 15 |
| Non-HDLc (mg/dl) | | | | | | |
| Baseline | 190 ± 35 | 194 ± 23 | 191 ± 31 | 184 ± 22 | 184 ± 29 | 185 ± 25 |
| End | 183 ± 30 | 197 ± 20 | 187 ± 27 | 171 ± 27* | 167 ± 30* | 170 ± 28*** |
| TG (mg/dl) | | | | | | |
| Baseline | 128 ± 73 | 109 ± 35 | 126 ± 65 | 97 ± 47 | 121 ± 78 | 104 ± 59 |
| End | 119 ± 71 | 100 ± 39 | 116 ± 64 | 103 ± 69 | 109 ± 68 | 103 ± 65 |
| Apo B/Apo A ratio | | | | | | |
| Baseline | 0.754 ± 0.163 | 0.864 ± 0.163 | 0.786 ± 0.163 | 0.702 ± 0.168 | 0.801 ± 0.211 | 0.733 ± 0.181 |
| End | 0.759 ± 0.174 | 0.817 ± 0.158 | 0.774 ± 0.167 | 0.644 ± 180** | 0.744 ± 0.168* | 0.669 ± 0.175*** |

Data are expressed as mean ± SD. Values of serum TG concentrations were normalized using a log transformation. * $p < 0.05$, **$p < 0.01$, ***$p < 0.001$ when compared using a paired Student's t-test for all subjects ($n = 34$ and 41 in diet and PS group, respectively), or a 2-factor measure repeated ANOVA (phenotype and time) followed by a paired Student's t-test.

Abbreviations: PS: phytosterols; TC: total cholesterol; LDLc: LDL cholesterol; HDLc: HDL cholesterol; TG: triglycerides; Apo: Apolipoprotein.

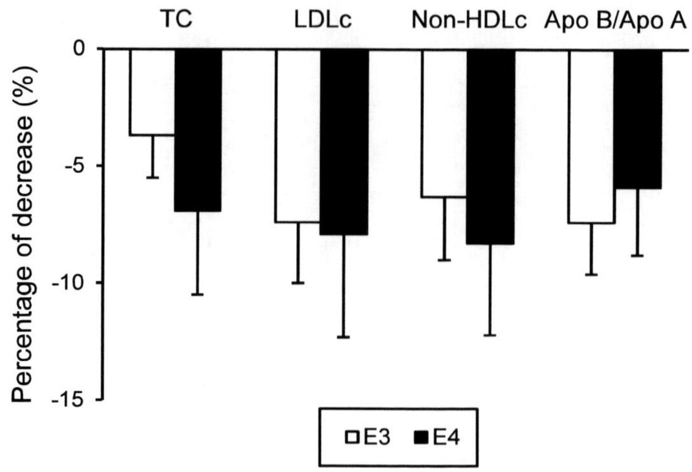


Data are represented as mean + SEM of 41 subjects for diet plus phytosterols group (n = 24 for E3 and n = 13 for E4). No differences between groups were found by unpaired Student t-test.

Abbreviations: TC: total cholesterol; LDLc: LDL cholesterol; HDLc: HDL cholesterol; Apo: Apolipoprotein.

Figure 5. Percentage of change in lipoprotein profile after intervention period in hypercholesterolemic subjects after addition of phytosterols to their diet according to ApoE phenotype.

**Table 5. Nutrient composition of a standard healthy diet and a free diet for hypercholesterolaemic patients**

| | Healthy Diet | Healthy Diet + PS | Free Diet + PS |
|---|---|---|---|
| N | 24 | 31 | 29 |
| Carbohydrates (%) | 51.6 ± 6.9 | 52.2 ± 7.4 | 48.5 ± 8.1 |
| Protein (%) | 18.5 ± 5.3 | 17.7 ± 6.5 | 17.3 ± 5.4 |
| Fat (%) | 29.9 ± 6.0[a] | 30.1 ± 5.7[a] | 34.2 ± 7.9[b] |
| Saturated fat (%) | 6.9 ± 2.6[a] | 6.7 ± 3.5[a] | 12.7 ± 3.7[b] |
| Polyunsaturated fat (%) | 6.1 ± 2.6 | 6.3 ± 2.2 | 6.1 ± 3.1 |
| Monounsaturated fat (%) | 16.9 ± 4.7 | 17.1 ± 4.1 | 15.4 ± 5.3 |
| Cholesterol (mg) | 193 ± 52[a] | 196 ± 67[a] | 268 ± 70[b] |

Average daily nutrient intake as percentage of total dietary caloric intake; 3-day food records (data collected on weekdays) and 24-hour diet recall.

Different superscript letters represent $p < 0.05$ when comparing the values from different treatments by one-way ANOVA. PS: phytosterols.

When we evaluated the efficacy of PS according to background diet, we recruited a total of eighty-four individuals with an age range of 18-76. The diet recalls and records revealed no deviations from the guidelines during the course of the study (Table 5). Men and women were distributed similarly among the three groups, with men constituting 25% of the healthy diet group, 29% of the healthy diet plus PS group and 35% of the free diet plus PS group. At onset of the study no significant differences were observed between the anthropometric parameters of patients receiving /not receiving a diet or beverages enriched with PS (data not shown).

After three months of PS therapy, patients on the free diet and on the healthy diet supplemented with PS exhibited a significant reduction in total cholesterol, LDLc, non-HDLc and Apo B/Apo A ratio, with a similar hypolipemic effect displayed in both groups (Table 6). In contrast, these lipid parameters were unaltered in patients undergoing dietary therapy only. No changes were found with respect to HDLc or TG in any of the groups (Table 6).

A standard healthy diet did not affect the levels of serum PS during the three-month monitoring period. However, intake of PS increased serum levels of campesterol and β-sitosterol to a similar extent independently of the type of diet followed (Table 7).

**Table 6. Lipoprotein profile of hypercholesterolemic subjects before and after addition of phyotosterols (PS) to a free or healthy diet**

| | Healthy Diet | Free Diet + PS | Healthy Diet + PS |
|---|---|---|---|
| TC (mg/dl) | | | |
| Baseline | 234 ± 21 | 230 ± 19 | 236 ± 18 |
| End | $233 \pm 26^{a}$ | $216 \pm 23^{b*}$ | $219 \pm 21^{b*}$ |
| LDLc (mg/dl) | | | |
| Baseline | 162 ± 15 | 158 ± 16 | 163 ± 16 |
| End | $162 \pm 25^{a}$ | $145 \pm 18^{b*}$ | $146 \pm 19^{b*}$ |
| HDLc (mg/dl) | | | |
| Baseline | 50 ± 13 | 52 ± 17 | 52 ± 14 |
| End | 50 ± 17 | 51 ± 19 | 54 ± 14 |
| TG (mg/dl) | | | |
| Baseline | 110 ± 62 | 101 ± 34 | 109 ± 63 |
| End | 110 ± 65 | 88 ± 50 | 95 ± 51 |
| non-HDLc (mg/dl) | | | |
| Baseline | 184 ± 21 | 178 ± 17 | 185 ± 20 |
| End | $184 \pm 25^{a}$ | $165 \pm 23^{b*}$ | $165 \pm 22^{b*}$ |
| Apo B/Apo A ratio | | | |
| Baseline | 0.770 ± 0.145 | 0.746 ± 0.164 | 0.721 ± 0.168 |
| End | $0.772 \pm 0.167^{a}$ | $0.674 \pm 0.158^{b*}$ | $0.635 \pm 0.132^{b*}$ |

Data are expressed as mean ± SD. Values of serum TG concentrations were normalized using a log transformation. Data were analysed by 2-factor repeated-measures ANOVA followed by paired Student's t-test and one way ANOVA to compare the effect of time for each treatment or the effect of treatment at different times, respectively. $*p < 0.05$ when compared using a paired Student's t-test. Different superscript letters represent mean $p < 0.05$ when comparing values from different treatments by one-way ANOVA.

Abbreviations: PS: phytosterols; TC: total cholesterol; LDLc: LDL cholesterol; HDLc: HDL cholesterol; TG: triglycerides; non-HDLc: non-HDL cholesterol; Apo: Apolipoprotein.

Serum concentration of liposoluble antioxidants (measured as carotenoid levels) in patients at the beginning and at the end of the trial are shown in Figure 6. Lutein and cryptoxanthin concentrations were similar in the three groups at baseline and at the end of the study, and showed no changes after the corresponding intervention (Figure 6). However, β -carotene was significantly increased (26.9%) by the standard healthy diet and was reduced by 21.0% by the free diet enriched with PS, whereas no change was observed in the group that followed a standard healthy diet with a low-fat milk enriched with PS (Figure 6). A moderate, though non-significant, increase of the other hydrocarbon carotenoid, lycopene, was observed in patients that followed the dietetic guidelines, whereas the difference was reduced significantly (22.8%) in patients that consumed PS without modifying their dietary pattern, and was maintained at baseline levels in the group that received the combined therapy (Figure 6). When we analyzed oxidative stress parameters, we found that LDL particles' susceptibility to oxidation was improved in those that followed the diet alone, as lag time in these subjects increased 4.6%, an effect that was not observed in the PS + diet group (Figure 7A). However, the enhancement of endogenous antioxidants observed as a result of adherence to the diet did not alter the total antioxidant status of the subjects, as can be seen in Figure 7B, or lipid peroxidation, an indicator of oxidative stress, as shown by Figure 7C.

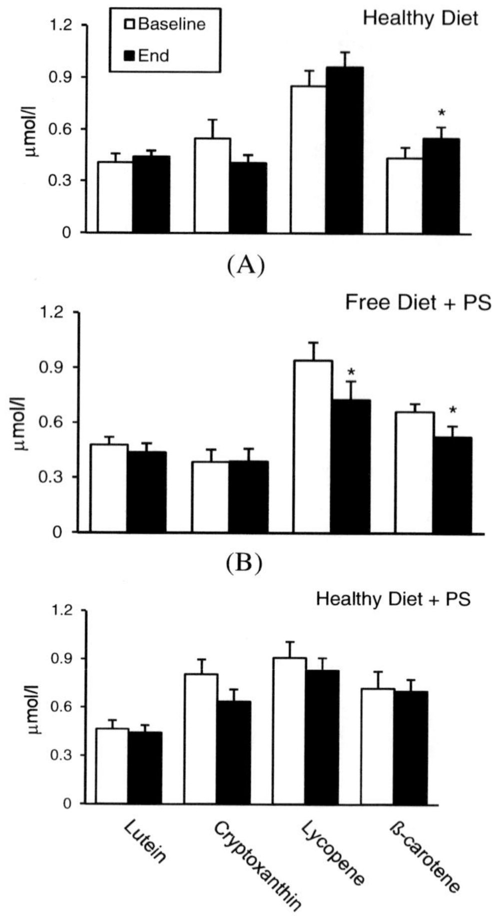


A. Levels of lutein, cryptoxanthin, lycopene and β-carotene in subjects who followed a standard healthy diet at baseline and after three months of intervention. Data are expressed as mean + SEM of 16 subjects.

B. Levels of lutein, cryptoxanthin, lycopene and β-carotene in subjects who followed a free diet complemented with phytosterols (PS) at baseline and after three months of intervention. Data are expressed as mean + SEM of 12 subjects.

C. Levels of lutein, cryptoxanthin, lycopene and β-carotene in subjects that followed a standard healthy diet complemented with phytosterols (PS) at baseline and after three months of intervention. Data are expressed as mean + SEM of 18 subjects.

*$p < 0{\cdot}05$ when compared using a paired Student t-test.

Figure 6. Concentration of carotenoids in the serum of hypercholesterolemic subjects at baseline and after three months of the corresponding intervention.

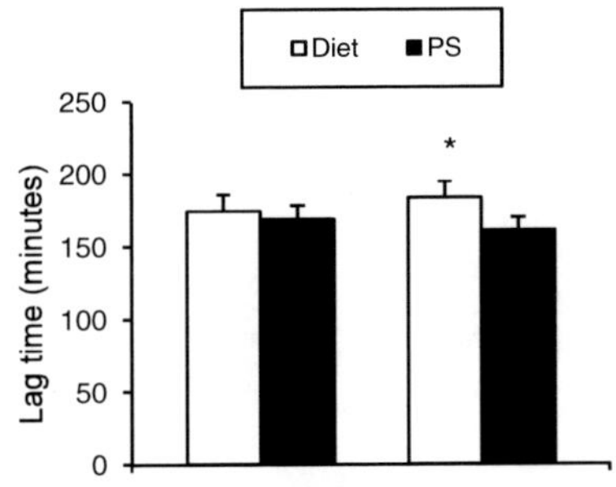


(A)

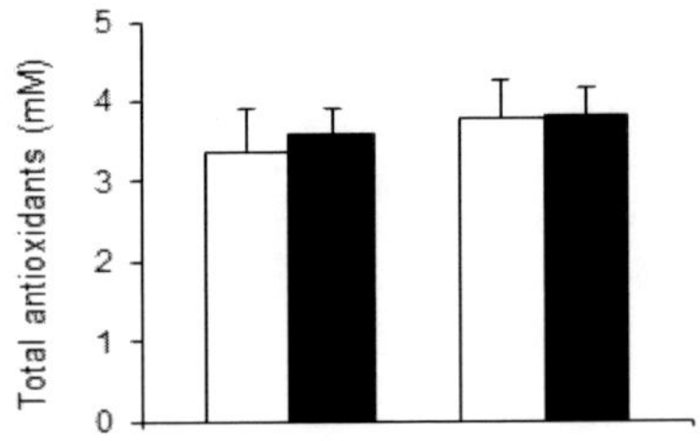


(B)

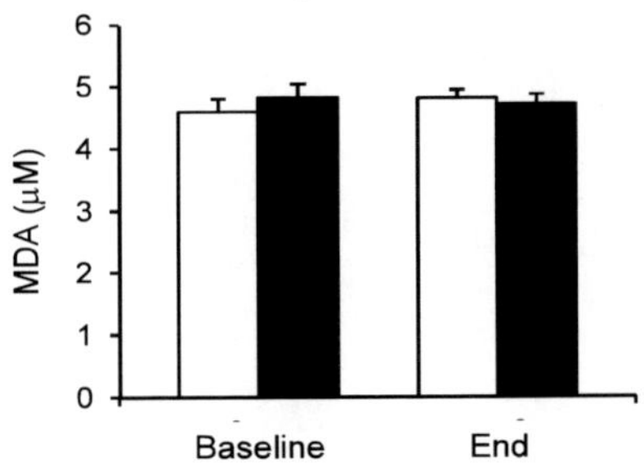


(C)

Panel A shows the susceptibility to oxidation in LDL particles represented as lag time.
Panel B shows global antioxidant status represented as total antioxidants.
Panel C shows lipid peroxidation represented as MDA.
Data are expressed as means ± SEM of 9 subjects for panel A, 12-16 subjects for panel B and 18 subjects for panel C. * $p < 0.05$ when compared using a 2-factor repeated-measures ANOVA followed by paired Student's t-test.
Abbreviations: PS: phytosterols; malondialdehyde: MDA.

Figure 7. Oxidative stress parameters in hypercholesterolemic subjects after addition of phytosterols (PS) to their diet.

After addition of PS to the diet, no significant additional changes were induced in BMI, waist perimeter or BP in any of the study groups (data not shown).

Enrichment of a healthy diet with PS induced a significant drop of 6.9% and 10.5% in total cholesterol and LDLc, respectively, in the moderate hypercholesterolemic population without MetS (Table 9 and Figure 8). This reduction was accompanied by a significant decrease of non-HDLc and a reduced cardiovascular risk, deduced from their lower Apo B/Apo A ratio (Table 9 and Figure 8). However, no changes were observed in HDLc or TG. Surprisingly, none of the abovementioned alterations was detected in subjects diagnosed as MetS (Table 9).

**Table 7. Concentration of phytosterols (PS) in serum at baseline and after three months of the corresponding intervention**

| | Healthy Diet | Free Diet + PS | Healthy Diet + PS |
|---|---|---|---|
| Campesterol (μmol/l) | | | |
| Baseline | 0.234 ± 0.082 | 0.231 ± 0.136 | 0.253 ± 0.142 |
| End | 0.233 ± 0.109[a] | 0.415 ± 0.189[b*] | 0.518 ± 0.432[b*] |
| β-Sitosterol (μmol/l) | | | |
| Baseline | 0.382 ± 0.218 | 0.400 ± 0.248 | 0.399 ± 0.216 |
| End | 0.340 ± 0.139[a] | 0.594 ± 0.246[b*] | 0.690 ± 0.493[b*] |

Data are expressed as mean ± SD of 16 subjects for healthy diet group, 12 for free diet + PS group and 18 for healthy diet + PS group. Data were analyzed by 2-factor repeated-measures ANOVA followed by paired Student's t-test and one way ANOVA to compare the within-group effect or between-group effect, respectively. *p < 0.05 when compared using a paired Student's t-test. Different superscript letters represent p < 0.05 when comparing values from different treatments by one-way ANOVA.

**Table 8. Anthropometric, lipoprotein profile and carbohydrate metabolism parameters of hypercholesterolemic subjects with or without metabolic syndrome at inclusion and after three months following a healthy diet**

| | With metabolic syndrome | | Without metabolic syndrome | |
|---|---|---|---|---|
| | Inclusion | 3 months diet (Baseline) | Inclusion | 3 months diet (Baseline) |
| Anthropometric parameters | | | | |
| Males/females (n) | 11/13 | ----------- | 10/14 | ----------- |
| Age (years) | 52.3 ± 10.5 | ----------- | 50.8 ± 13.8 | ----------- |
| BMI (Kg/m$^2$) | 32.2 ± 4.8 | 31.2 ± 4.3** | 30.0 ± 3.4 | 28.6 ± 3.7** |
| Waist (cm) | 104 ± 12 | 100 ± 11** | 92 ± 9## | 92 ± 9 |
| Systolic BP (mmHg) | 142 ± 14 | 138 ± 17 | 132 ± 21 | 130 ± 14 |
| Diastolic BP (mmHg) | 91 ± 8 | 94 ± 8 | 82 ± 12# | 81 ± 9 |
| Lipoprotein profile | | | | |
| TC (mg/dl) | 237 ± 22 | 231 ± 24 | 253 ± 22# | 239 ± 26** |
| LDLc (mg/dl) | 158 ± 20 | 157 ± 21 | 177 ± 14## | 168 ± 21* |
| HDLc (mg/dl) | 46 ± 10 | 44 ± 12 | 51 ± 10 | 48 ± 12 |
| Non-HDLc (mg/dl) | 191 ± 22 | 187 ± 23 | 202 ± 19 | 191 ± 27* |
| TG (mg/dl) | 151 ± 61 | 147 ± 67 | 138 ± 48# | 111 ± 57 |
| Apo B/Apo A ratio | 0.798 ± 0.220 | 0.800 ± 0.198 | 0.755 ± 0.153 | 0.731 ± 0.128 |
| Diameter LDL (nm) | 25.9 ± 0.8 | 25.8 ± 0.6 | 26.3± 0.5# | 26.3 ± 0.9 |
| Carbohydrate metabolism | | | | |
| Glucose (mg/dl) | 107 ± 11 | 103 ± 14* | 98 ± 14# | 99 ± 12 |
| Insulin (μU/ml) | 12.5 ± 6.0 | 11.3 ± 5.3 | 10.1 ± 4.9 | 9.9 ± 5.3 |
| HOMA-IR index | 3.38 ± 1.90 | 2.86 ± 1.47 | 2.51 ± 1.48 | 2.45 ± 1.46 |

Data are expressed as mean ± SD of 24 subjects for each study. Values of serum TG concentrations were normalised using a log transformation. *p < 0.05 and **p < 0.01, after the first three months of the study, values were compared with baseline levels using a paired Student's t-test. #p < 0.05 and ##p < 0.01, baseline characteristic of subjects with or without metabolic syndrome were compared by means of an unpaired Student's t-test.

Abbreviations: TC: total cholesterol; LDLc: LDL cholesterol; HDLc: HDL cholesterol; non-HDLc: non-HDL cholesterol; TG: triglycerides; Apo: Apolipoprotein.

Finally, we evaluated whether the lower absorption of cholesterol observed in MetS patients interfered with PS' mechanism of action, thus reducing their efficacy as hypocholesterolemic agents. Our results showed that, at baseline, MetS patients presented a higher waist perimeter, diastolic BP, TG and glucose than subjects without MetS, and that they also exhibited lower serum levels of total cholesterol and LDLc (Table 8).

The dietary intervention induced a significant decrease of BMI in both populations and a reduction in waist perimeter and glucose levels only in subjects with MetS. In contrast, in the non-MetS population, the healthy diet was effective in reducing total cholesterol, LDLc, and non-HDLc, as shown in Table 8.

Baseline serum PS concentrations, measured as campesterol and β-sitosterol, were found to be significantly higher in subjects without MetS than in those diagnosed as having MetS (Figure 9). Furthermore, addition of PS to the dietary guidelines provoked a noticeable increase in serum concentration of campesterol and β-sitosterol, although a statistical significance was reached only in the population without MetS (Table 10). Similar observations were obtained when serum PS concentrations were normalized to cholesterol levels (data not shown).

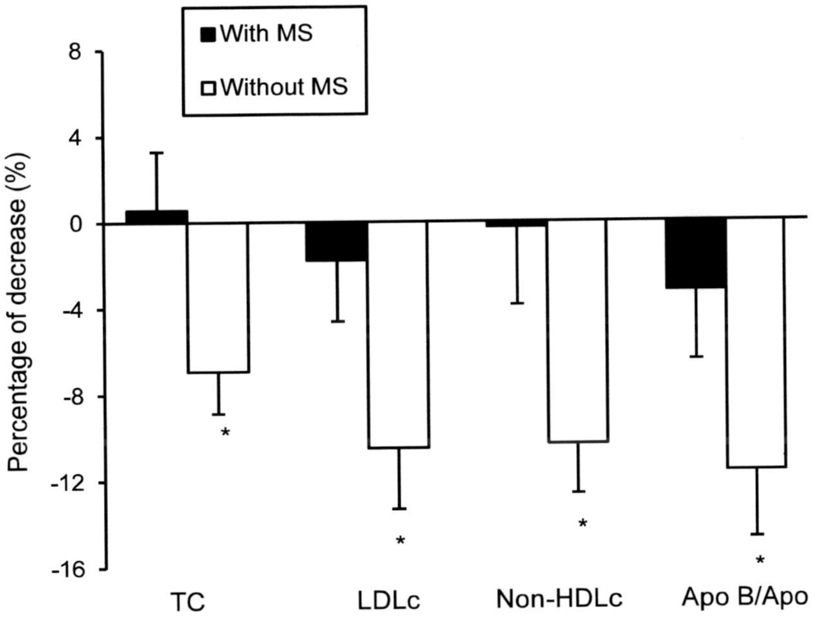


Data are represented as mean + SEM of 14 subjects of each study. *$p < 0.05$, when compared by means of an unpaired Student's t-test.

Abbreviations: PS: phytosterols; TC: total cholesterol; LDLc: LDL cholesterol; non-HDLc: non-HDL cholesterol; Apo: Apolipoprotein.

Figure 8. Percentage of change in lipid parameters after receiving diet plus phytosterols (PS) in hypercholesterolemic subjects with or without metabolic syndrome.

**Table 9. Lipoprotein profile and carbohydrate metabolism parameters of hypercholesterolemic subjects with or without metabolic syndrome after addition of phytosterols (PS) to their diet**

| | With metabolic syndrome | | Without metabolic syndrome | |
|---|---|---|---|---|
| | Baseline | End | Baseline | End |
| Males/Females (n) | | | | |
| Diet | 3/7 | | 4/6 | |
| PS | 5/9 | | 6/8 | |
| TC (mg/dl) | | | | |
| Diet | 234 ± 25 | 230 ± 17 | 235 ± 34 | 235 ± 31 |
| PS | 228 ± 23 | 229 ± 33 | 239 ± 26 | 225 ± 23** |
| LDLc (mg/dl) | | | | |
| Diet | 157 ± 25 | 154 ± 17 | 163 ± 23 | 162 ± 28 |
| PS | 158 ± 18 | 154 ± 22 | 172 ± 19 | 153 ± 18** |
| HDLc (mg/dl) | | | | |
| Diet | 45 ± 7 | 45 ± 8 | 50 ± 14 | 52 ± 12 |
| PS | 43 ± 14 | 44 ± 15 | 47 ± 11 | 49 ± 10 |
| Non-HDLc (mg/dl) | | | | |
| Diet | 189 ± 23 | 185 ± 15 | 185 ± 30 | 182 ± 31 |
| PS | 185 ± 23 | 185 ± 32 | 192 ± 22 | 176 ± 21*** |
| TG (mg/dl) | | | | |
| Diet | 162 ± 58 | 152 ± 66 | 111 ± 79 | 105 ± 78 |
| PS | 138 ± 73 | 150 ± 82 | 110 ± 38 | 101 ± 38 |
| Apo B/Apo A ratio | | | | |
| Diet | 0.795 ± 0.107 | 0.799 ± 0.116 | 0.728 ± 0.160 | 0.760 ± 0.219 |
| PS | 0.803 ± 0.247 | 0.769 ± 0.228 | 0.733 ± 0.104 | 0.646 ± 0.108** |
| Diameter LDL (nm) | | | | |
| Diet | 25.8 ± 0.6 | 25.5 ± 0.7 | 26.2 ± 0.6 | 26.5 ± 0.8 |
| PS | 25.8 ± 0.6 | 25.8 ± 0.9 | 26.3 ± 1.0 | 26.5 ± 0.9 |
| Glucose (mg/dl) | | | | |
| Diet | 109 ± 14 | 107 ± 20 | 100 ± 17 | 99 ± 18 |
| PS | 99 ± 13 | 96 ± 7 | 98 ± 8 | 96 ± 12 |
| Insulin (μU/ml) | | | | |
| Diet | 13.1 ± 6.5 | 9.9 ± 2.8 | 10.2 ± 8.7 | 11.5 ± 9.4 |
| PS | 10.3 ± 4.6 | 12.7 ± 4.2 | 11.3 ± 5.7 | 10.1 ± 5.0 |
| HOMA-IR index | | | | |
| Diet | 3.39 ± 1.54 | 2.51 ± 0.76 | 1.94 ± 1.16 | 2.14 ± 1.50 |
| PS | 2.60 ± 1.41 | 3.05 ± 1.12 | 2.77 ± 1.58 | 2.48 ± 1.56 |

Data are expressed as mean ± SD. After the first three months of dietary therapy, both subjects with and without metabolic syndrome were randomized into two groups for a further three-month period. In the metabolic syndrome group, 10 subjects followed dietary recommendations (Diet) and 14 subjects enriched their diet with 2g/day of PS. In the group without metabolic syndrome, 10 subjects followed dietary recommendations (Diet) and 14 subjects enriched their diet with 2g/day of PS. Values of serum TG concentrations were normalised using a log transformation. *$p < 0.05$, **$p < 0.01$, ***$p < 0.001$ when compared by 2-factor repeated-measures ANOVA followed by a paired Student's t-test after addition of PS.

Abbreviations: PS: phytosterols; TC: total cholesterol; LDLc: LDL cholesterol; HDLc: HDL cholesterol; non-HDLc: non-HDL cholesterol; TG: triglycerides; Apo: Apolipoprotein.

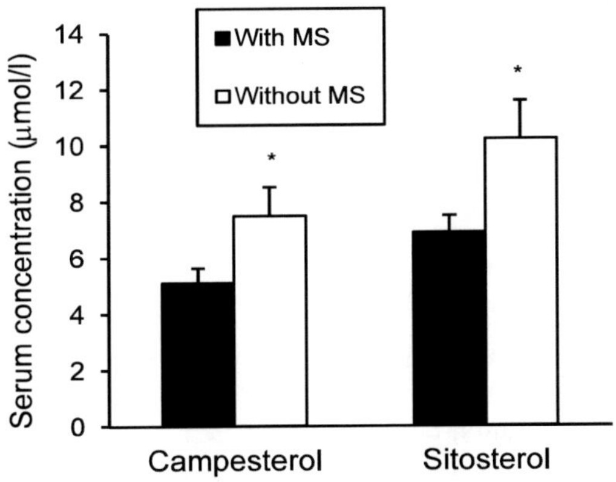


Data are represented as mean + SEM of 12 subjects with metabolic syndrome and 14 subjects without metabolic syndrome. *p < 0.05, when compared by means of an unpaired Student's t-test.

Figure 9. Baseline serum concentration of phytosterols (PS) in hypercholesterolemic subjects with or without metabolic syndrome.

**Table 10. Serum concentration of campesterol and β-sitosterol before and after addition of phytosterols (PS) to the diet of hypercholesterolemic subjects with or without metabolic syndrome**

| | | Metabolic Syndrome | | Without metabolic syndrome | |
|---|---|---|---|---|---|
| | | **Diet (n = 7)** | **PS (n = 5)** | **Diet (n = 7)** | **PS (n = 6)** |
| Campesterol (μmol/l) | Inclusion | 0.203 ± 0.087 | 0.207 ± 0.057 | 0.281 ± 0.135 | 0.265 ± 0.213 |
| | End | 0.262 ± 0.132 | 0.362 ± 0.226 | 0.199 ± 0.085 | 0.427 ± 0.281* |
| β-Sitosterol (μmol/l) | Inclusion | 0.291 ± 0.095 | 0.280 ± 0.071 | 0.331 ± 0.109 | 0.407 ± 0.278 |
| | End | 0.370 ± 0.188 | 0.469 ± 0.273 | 0.278 ± 0.068 | 0.650 ± 0.364** |

Data are expressed as mean ± SD. Serum concentrations were analysed at the beginning of the study (Inclusion) and after 6 months (End) following the diet or a combination of diet + phytosterols (PS). *p < 0.05 when compared by means of a 2-factor repeated-measures ANOVA followed by a paired Student's t-test.

## DISCUSSION

Since the daily consumption of PS in a standard healthy diet is lower than 2 g, which is considered the amount necessary for it to exert its therapeutic effects, we evaluated whether the intake of a PS-enriched beverage would exert an hypolipidemic effect independently of basal LDLc concentration in a general population. Our results showed that the study population was highly heterogeneous (it included normocholesterolemic patients and moderate hypercholesterolemic patients), with a total cholesterol interval between 138 and 294 mg/dl and LDLc between 78 and 190 mg/dl. In the whole population, addition of PS to the subjects' usual diet induced a significant decrease in total cholesterol and LDLc of 5.3% and 6.5%, respectively, which is in accordance with the findings of previous studies (Clifton et al. 2004b; Noakes et al. 2005).

In line with ATPIII, we stratified patients by tertiles of LDLc in order to evaluate the hypolipidemic effect of the PS-enriched beverage. A significant decrease of 7.3% was found in patients in the third tertitle, in which LDLc was higher than 154 mg/dl. These results are in accordance with those of Seppo et al. (2007), who concluded that PS-enriched products decreased LDLc by 8.4 % in subjects with LDLc over 135 mg/dl when consumed for five weeks. Thus, our data suggest that reduction of LDLc is more pronounced in subjects with higher baseline concentrations of cholesterol.

Patients with hypercholesterolemia are advised first and foremost to limit their intake of total and saturated fats, in particular fats obtained from dairy products (Expert panel on detection, evaluation, and treatment of high blood cholesterol in adults, 2001). For this reason, previous to PS consumption, patients were submitted to a three month run-in period in which a healthy diet was recommended. Thus, one of the strengths of this randomized parallel study is the fact that participants received dietary guidance by an experienced dietician throughout the trial. Our results show that three months of healthy diet previous to randomization reduced total cholesterol, LDLc and non-HDLc. No additional decrease in serum total cholesterol or LDLc concentration was recorded after this period. The approximate 5% decrease in LDLc observed is in accordance with that estimated by Hunninghake et al. (1993). In fact, Bae et al. (1991) also reported a similar pattern of response after 6 weeks of diet intervention, but observed no further significant decrease in the following 12 weeks.

Importantly, and in line with the results of several studies (Katan et al., 2003; Law, 2000), we have demonstrated that the ingestion of approximately 2 g/day PS is associated with a 7.0–10.0% decrease in LDLc levels. PS are included in the National Cholesterol Education Program (NCEP) (Expert panel on detection, evaluation, and treatment of high blood cholesterol in adults, 2001) and the American Heart Association/American College of Cardiology (De Backer et al., 2004) guidelines for secondary prevention of CHD. For each percentage of LDLc reduction, the risk of CHD disease is reduced by 2–3%. In the second intervention phase of our trial, ingestion of low-fat milk enriched with PS reduced total cholesterol concentration and LDLc by 6.4% and 9.9%, respectively, whereas dietary measures alone reduced these values by only 2.1% and 1.8%, respectively. Therefore, it would appear that, in moderately hypercholesterolemic subjects, the combination of dietary therapy and PS-enriched low-fat milk reduces the need for cholesterol-lowering drug therapy.

Numerous studies have shown that a predominance of small and dense LDL particles is associated with an increased cardiovascular risk (Packard, 2006). In fact, small and dense LDL particles exhibit a reduced affinity for the LDL receptor (Nigon et al., 1991) and a marked susceptibility to oxidation, inducing alterations in endothelial function. Our results support the idea that a decrease in total cholesterol and LDLc does not have beneficial effects on LDL size, since Apo B concentration remained unaltered in our subjects. They suggest that PS do not affect LDL particle size as part of their mechanism of action, which is in accordance with those of previous studies (Matvienko et al., 2002; Charest et al., 2005; Gigleux et al., 2007). It could be argued that this response is to be expected, as changes in TG are known to be the main cause of the modification of LDL particle size and PS therapy does not alter TG serum levels. The fact that others have described an increase in LDL diameter following therapy combining PS with soluble fibre or culinary oils (St-Onge et al., 2003; Shrestha et al.,2007), or sterols with exercise (Varady et al., 2005), suggest that the positive effect on LDL size is induced by other stimuli that modify TG concentration to a greater extent than PS.

Despite the fact that most previous studies confirm the hypolipidemic effect of PS, it is not clear whether this effect is influenced or not by the Apo E genotype. Several studies have investigated plasma lipid responses to PS supplementation in the context of the ApoE locus, and, although most have shown that sterols and stanols lower LDLc irrespectively of ApoE alleles (Kempen et al., 1991; Geelen et al., 2002; Ishiwata et al., 2002; Lottenberg et al., 2002; Plat and Mensink, 2002; Tammi et al., 2002), one recent report concluded that PS therapy was of little value to E4 subjects with hypercholesterolemia owing to the lack of significant benefits in relation to total cholesterol and LDLc concentrations (Sanchez-Muñiz et al., 2009). Our results are in accordance with those of previous studies in which PS therapy was not influenced by Apo E genotype. Among said studies, several were carried out in normocholesterolemic subjects (Kempen et al., 1991; Geelen et al., 2002; Ishiwata et al., 2002; Plat and Mensink, 2002; Tammi et al., 2002) and others were performed in a hypercholesterolemic population (Lottenberg et al., 2002; Sanchez-Muñiz et al., 2009) with different doses and different types of PS (free sterols, and stanols or stanol esters). In all cases, with the exception of the study by Ordovas (2009), non-significant variations were observed among the lipoprotein responses of individuals according to their Apo E phenotype (Sanchez-Muñiz et al., 2009).

As potential extrinsic factors modulating the efficacy of PS, we investigated the role of dietary cholesterol and saturated fat intake. As far as we know, this is the first human clinical study to compare the effects of esterified PS administered as part of a diet in which the levels of saturated fats and cholesterol intake were significantly different. Moreover, we employed a three-arm, randomised, parallel study design.

In line with previous studies (Katan et al., 2003, Abumweis et al., 2008, Demonty et al., 2009), we observed that ingestion of approximately 2 g of PS per day was associated with a 7.0–10.0% reduction in LDLc levels. In addition, we have seen that 3-month ingestion of low-fat milk enriched with PS produced similar reductions in total cholesterol and LDLc concentrations (5.5 and 6.7% of total cholesterol and 7.0 and 9.6% of LDLc in free and healthy diets, respectively). Only a few studies have analysed the efficacy of PS in relation to dietary cholesterol intake. Jones (2015) did not observe variations in the effectiveness of free sterols when a single dose was administered in the morning, which is in accordance with our findings (Kassis et al., 2008). On the other hand, other studies have shown that low levels of dietary cholesterol attenuate the efficacy of a low-dose PS therapy, or have no effect at all (Denke et al., 1995, Mussner et al., 2002). Therefore, it would appear that, in moderately hypercholesterolemic subjects, the efficacy of PS with respect to lipoprotein profile is not influenced by the dietary composition of saturated fat or cholesterol intake, at least under the conditions of our experimental design. This finding could be of special relevance, since these patients are likely to continue their free diet, which can be supplemented with PS-enriched food.

Measurement of serum PS has revealed that campesterol and β-sitosterol concentrations increase with consumption of PS, independently of diet, suggesting that dietary cholesterol has no effect on the absorption of PS. Due to cholesterol being absorbed via the same pathway as PS (Trautwein et al., 2003), we expected the stronger competition due to a higher cholesterol intake to reduce the amount of PS in circulation. However, it has been shown that exogenous cholesterol makes a very slight contribution to the level of cholesterol in the blood, reaching approximately 20% that of endogenously synthesised cholesterol. Moreover, the absorption of cholesterol and PS varies in the different parts of the intestine (Trautwein et al., 2003). Therefore, a change in dietary cholesterol is unlikely to alter PS bioavailability.

It is worth noting that campesterol seems to be absorbed to a greater degree than β-sitosterol; though the level of the former is about four times less than that of the latter in milk, its increase in serum is approximately twice as high. In addition, clearance of campesterol is lower than that of β-sitosterol, indicating that the two types of PS possess different kinetics (Sudhop et al., 2002).

The precise mechanism by which PS reduce plasma carotenoid concentration is not well defined, but it has been suggested that their incorporation into the mixed micelles (Brufau et al., 2008) and their variable bioavailability, which is influenced by many dietary and physiological factors (Yonekura and Nagao, 2007), could explain the conflicting results reported in the literature.

In our study, β-carotene and lycopene serum concentrations dropped following PS intake when the patients followed their regular diet, whereas oxygenated carotenoid levels (lutein and β-cryptoxanthin) were not modified. This may be due to the distribution of these antioxidants in the mixed micelles (surface or core). It is thought that apolar hydrocarbon carotenoids (β -carotene and lycopene) are solubilised in the core of mixed micelles, whereas oxygenated carotenoids are located mainly on the surface (Borel et al., 1996). This suggests that PS replace, not only cholesterol from the core of the mixed micelles, but also other compounds present in the micelles. Most reports associate intake of PS with a decrease in the levels of β-carotene (Colgan et al., 2004; Judd et al., 2002; Hansel et al., 2007; Noakes et al., 2005; Polagruto et al., 2006), whereas an additional daily serving of fruits and vegetables (with high levels of carotenoids) or products enriched with both carotenoids and PS ameliorates any reductions in plasma carotenoid concentration induced by PS (Brufau et al., 2004, Noakes et al., 2002; Clifton et al., 2004a; Colgan et al., 2004). This is compatible with the present results, and reinforces the idea that the enrichment of diet with fruits and vegetables has a positive effect on antioxidant levels.

In this sense, we also analysed the effect of daily intake of PS plus a healthy diet on oxidative stress parameters due to the involvement of PS in the development of atherosclerosis and other chronic pathologies (Victor and Rocha, 2007). It has been suggested that even a minor decrease in β-carotene concentration can be proatherogenic (D'Odorico et al., 2000). However, antioxidants include numerous substances or enzymes and liposoluble vitamins that may interact synergistically. Analysis of one component independently of the others ignores a possible combined action, and, therefore, does not allow systemic oxidative stress to be accurately estimated. For that reason, we evaluated the effect of PS on total antioxidant status, on the capacity of in vitro antioxidants to protect LDL particles against oxidation, and on lipid peroxidation. Interestingly, in vitro LDL oxidation resistance was greater among subjects who did not receive the PS supplement, in accordance with the increase in the concentration of liposoluble vitamins (β-carotene) associated with healthy diets enriched with fruits and vegetables. There is no consistently reported effect of diet on the susceptibility of LDL to oxidation. In one previous study, a vegetarian diet improved lag time to a similar extent as that achieved in our study (Korpela et al., 1999), whereas another report obtained no effect with a low-fat diet enriched with vegetable oils (Schwab et al., 1998). In our study, addition of PS did not alter the susceptibility of LDL to oxidation, as reported in a previous publication (Shrestha et al., 2007).

Despite changes in the concentration of serum liposoluble antioxidants and the capacity of in vitro antioxidants to protect LDL particles against oxidation, we detected no changes in serum total antioxidant status or oxidative stress (assessed as lipid peroxidation). There is a scarcity of evidence regarding the effect of PS on markers of lipid peroxidation levels, and the results that do exist are controversial. Studies spanning relatively short periods of time (4–8 weeks) have reported a decrease in lipid peroxidation parameters (Chan et al., 2007, Brufau et al., 2004, Mannarino et al., 2009), whereas no changes were observed when PS were consumed over 16 weeks (De Jong et al., 2008), in line with our data.

Finally, a human model of low absorption of cholesterol (MetS) was chosen to evaluate the efficacy of PS. In contrast with the results obtained in hypercholesterolemic patients, no modification was detected in any of the lipoprotein profile parameters in MetS patients after three months adherence to a healthy diet. Studies have shown that LDLc response to diets low in saturated fatty acids (Lefevre et al., 2005) or to high dietary cholesterol (Knopp et al., 2003) is much less pronounced in obese insulin resistant subjects than in lean insulin-sensitive individuals. The beneficial effect of a 3-month healthy diet on lipid profile in a hypercholesterolemic population without MetS has been discussed above.

It is well established that PS lower intestinal cholesterol absorption by 30-40%, (Ostlund, 2007), resulting in a significant reduction of serum LDLc concentration of about 10% when administered at a dose of 2 g/day (Demontly et al., 2009). In fact, in a non-MetS population, we achieved a significant reduction of total cholesterol, LDLc and non-HDLc after adding PS to a healthy diet, which represented a diminishing of cardiovascular risk factors among these patients. However, the potential effect of daily consumption of PS on MetS remains unknown. To our knowledge, only two studies have addressed this topic, and both employed small sample sizes (less than 10 subjects per group) (Ooi et al., 2007, Plat et al., 2009). Their findings are controversial; for instance, Plat et al. (2009) found a reduction of non-HDLc and TG in patients with overt hypercholesterolaemia and hypertriglyceridemia after 9 weeks of daily consumption of PS. They did not evaluate LDLc, despite the fact that PS specifically reduces this lipoprotein, and reported that Apo B remained unaltered. In contrast, another study carried out in men with MetS did not identify any real quantitative benefit of PS supplementation with respect to lipoprotein profile (Ooi et al., 2007), which is in accordance with the present findings. Also in line with our results, a previous study showed that addition of fibre -whose hypolipemic effect induces a 10-15% reduction of LDLc- to the diet of a hypercholesterolemic population with MetS failed to reduce serum total or LDLc after 12 weeks (Hallikainen et al., 2006) The effect of soluble fibre as a hypocholesterolemic agent is largely based on its ability to increase faecal bile acid excretion and to inhibit cholesterol absorption from the intestine (Fernandez, 2001), the latter of which is a mechanism of action common to both soluble fibre and PS.

There is strong evidence that the efficiency of cholesterol absorption is lower and cholesterol synthesis higher in insulin-resistant states such as MetS (Simonen et al., 2000, Pihlajamäki et al., 2004, Ooi et al., 2009). Although we did not measure cholesterol synthesis, our results indicate a low intestinal absorption of cholesterol in MetS patients. At baseline, cholesterol and LDLc levels were lower in the MetS population than in patients without MetS, despite the two groups being matched in BMI, age and sex. Baseline serum levels of campesterol and β-sitosterol -markers of cholesterol absorption- were also reduced in the MetS population. Adherence to a diet in which saturated fat and cholesterol intake were specifically decreased did not reduce serum cholesterol and LDLc levels in MetS patients, in contrast to that observed in the population without MetS. Finally, supplementation with PS did not induce the expected hypolipemic effect, probably due to a reduced absorption rate of PS. These findings could be explained by an enhanced cholesterol flux through the liver, which would down-regulate LDL receptors and make them refractory to additional regulation by dietary fatty acid changes, while a decreased cholesterol flux though enterocytes would lessen both the cholesterol-raising response to dietary cholesterol and the cholesterol-lowering effect of PS. In other words, dietary PS supplementation is of special benefit to people with high intestinal cholesterol absorption and low endogenous cholesterol synthesis (Ascaso et al., 2007). In addition to serum lipoprotein concentrations, markers related to the initiation and progression of atherosclerosis are also valid predictors of future cardiovascular diseases. In relation to whether or not PS affect LDL particle size, proinflammatory and prothrombotic states in non-MetS populations have been the subject of very little study, and even less in MetS subjects. The preponderance of small and dense LDL particles is one of the main causes of the initiation of atherosclerosis. In fact, these particles exhibit a reduced affinity for the LDL receptor and a marked susceptibility to oxidation, inducing alterations in endothelial function. Although these particles are smaller in MetS patients than in non-MetS subjects, our results suggest that PS did not affect LDL particle size in either population, which is in accordance with the findings of previous studies (Charest et al., 2005, Gigleux et al., 2007).

In summary, our results confirm and extend the therapeutic effect of regular consumption of PS in subjects with moderate hypercholesterolemia. This effect is potentiated by dietary measures that reduce the lipoprotein-mediated risk of CVD. Dietary composition of saturated fat or dietary cholesterol, on the other hand, does not have a bearing on the efficacy of PS. We also show that this hypercholesterolemic effect is not mediated by responses specific to Apo E phenotype or by smaller LDL particle size. Our results show that a decrease in carotenoids associated with PS consumption does not lead to global impairment of antioxidative defences or an enhancement of oxidative stress. Surprisingly, our MetS patients did not respond to the PS therapy, which appears to be a result of the reduced intestinal cholesterol absorption that characterises these patients, and which is likely to interfere with PS' mechanism of action.

## References

Abumweis, SS; Barake, R; Jones, PJ. Plant sterols/stanols as cholesterol lowering agents: A meta-analysis of randomized controlled trials. *Food Nutr Res*, 2008, 52.doi:10.3402/fnr.520.1811.

Abumweis, SS; Vanstone, CA; Ebine, N; Kassis, A; Ausman, LM; Jones, PJ; et al. Intake of a single morning dose of standard and novel plant sterol preparations for 4 weeks does not dramatically affect plasma lipid concentrations in humans. *J Nutr*, 2006, 136(4), 1012-1016.

Allain, CC; Poon, LS; Chan, CS; Richmond, W; Fu, PC. Enzymatic determination of total serum cholesterol. *Clin Chem,* 1974, 20(4), 470-475.

Ascaso, J; Santos, PG; Mijares, AH; Rojas, AM; Masana, L; Millan, J; et al. Management of dyslipidemia in the metabolic syndrome: recommendations of the Spanish HDL-Forum. *Am J Cardiovasc Drugs*, 2007, 7(1), 39-58.

Awika, JM; Rooney, LW. Sorghum phytochemicals and their potential impact on human health. *Phytochemistry*, 2004, 65(9), 1199-1221.

Bae, CY; Keenan, JM; Wenz, J; McCaffrey, DJ. A clinical trial of the American Heart Association step one diet for treatment of hypercholesterolemia. *J Fam Pract*, 1991, 33(3), 249-254.

Borel, P; Grolier, P; Armand, M; Partier, A; Lafont, H; Lairon, D; et al. Carotenoids in biological emulsions: solubility, surface-to-core distribution, and release from lipid droplets. *J Lipid Res*, 1996, 37(2), 250-261.

Brufau, G; Canela, MA; Rafecas, M. Phytosterols: physiologic and metabolic aspects related to cholesterol-lowering properties. *Nutr Res*, 2008, 28(4), 217-225.

Brufau, G; Quilez, J; Canela, MA; Salas-Salvado, J; Bullo, MM; Rafecas, M. Evaluation of lipid oxidation after ingestion of bakery products enriched with phytosterols, beta-carotene and alpha-tocopherol. *Clin Nutr*, 2004, 23(6), 1390-1397.

Bucolo, G; David, H. Quantitative determination of serum triglycerides by the use of enzymes. *Clin Chem*, 1973, 19(5), 476-482.

Calpe-Berdiel, L; Escola-Gil, JC; Blanco-Vaca, F. New insights into the molecular actions of plant sterols and stanols in cholesterol metabolism. *Atherosclerosis*, 2009, 203(1), 18-31.

Chan, YM; Demonty, I; Pelled, D; Jones, PJ. Olive oil containing olive oil fatty acid esters of plant sterols and dietary diacylglycerol reduces low-density lipoprotein cholesterol and decreases the tendency for peroxidation in hypercholesterolaemic subjects. *Br J Nutr*, 2007, 98(3), 563-570.

Charest, A; Vanstone, C; St-Onge, MP; Parson, W; Jones, PJ; Lamarche, B. Phytosterols in nonfat and low-fat beverages have no impact on the LDL size phenotype. *Eur J Clin Nutr*, 2005, 59(6), 801-804.

Cleghorn, CL; Skeaff, CM; Mann, J; Chisholm, A. Plant sterol-enriched spread enhances the cholesterol-lowering potential of a fat-reduced diet. *Eur J Clin Nutr*, 2003, 57(1), 170-176.

Clifton, PM; Noakes, M; Ross, D; Fassoulakis, A; Cehun, M; Nestel, P. High dietary intake of phytosterol esters decreases carotenoids and increases plasma plant sterol levels with no additional cholesterol lowering. *J Lipid Res*, 2004a, 45(8), 1493-1499.

Clifton, PM; Noakes, M; Sullivan, D; Erichsen, N; Ross, D; Annison, G; et al. Cholesterol-lowering effects of plant sterol esters differ in milk, yoghurt, bread and cereal. *Eur J Clin Nutr*, 2004b, 58(3), 503-509.

Colgan, HA; Floyd, S; Noone, EJ; Gibney, MJ; Roche, HM. Increased intake of fruit and vegetables and a low-fat diet, with and without low-fat plant sterol-enriched spread consumption: effects on plasma lipoprotein and carotenoid metabolism. *J Hum Nutr Diet*, 2004, 17(6), 561-9.

Davidson, MH; Kong, JC; Drennan, KB; Story, K; Anderson, GH. Efficacy of the National Cholesterol Education Program Step I Diet. A randomized trial incorporating quick-service foods. *Arch Intern Med*, 1996, 156(3), 305-312.

De Backer, G; Ambrosioni, E; Borch-Johnsen, K; Brotons, C; Cifkova, R; Dallongeville, J; et al. European guidelines on cardiovascular disease prevention in clinical practice. Third Joint Task Force of European and other Societies on Cardiovascular Disease Prevention in Clinical Practice (constituted by representatives of eight societies and by invited experts). *Atherosclerosis*, 2004, 173(2), 381-391.

De Jong, AD; Plat, J; Bast, A; Godschalk, RW; Basu, S; Mensink, RP. Effects of plant sterol and stanol ester consumption on lipid metabolism, antioxidant status and markers of oxidative stress, endothelial function and low-grade inflammation in patients on current statin treatment. *Eur J Clin Nutr*, 2008, 62(2), 263-273.

Demonty, I; Ras, RT; Knaap, HCvd; Duchateau, GS; Meijer, L; Zock, PL; et al. Continuous dose-response relationship of the LDL-cholesterol-lowering effect of phytosterol intake. *J Nutr*, 2009, 139(2), 271-284.

Denke, MA. Lack of efficacy of low-dose sitostanol therapy as an adjunct to a cholesterol-lowering diet in men with moderate hypercholesterolemia. *Am J Clin Nutr*, 1995, 61(2), 392-396.

Devaraj, S; Autret, BC; Jialal, I. Reduced-calorie orange juice beverage with plant sterols lowers C-reactive protein concentrations and improves the lipid profile in human volunteers. *Am J Clin Nutr*, 2006, 84(4), 756-761.

Devaraj, S; Jialal, I; Vega-Lopez, S. Plant sterol-fortified orange juice effectively lowers cholesterol levels in mildly hypercholesterolemic healthy individuals. *Arterioscler Thromb Vasc Biol*, 2004, 24(3), e25-8.

D'Odorico, A; Martines, D; Kiechl, S; Egger, G; Oberhollenzer, F; Bonvicini, P; et al. High plasma levels of alpha- and beta-carotene are associated with a lower risk of atherosclerosis: results from the Bruneck study. *Atherosclerosis*, 2000, 153(1), 231-239.

Esterbauer, H; Striegl, G; Puhl, H; Rotheneder, M. Continuous monitoring of in vitro oxidation of human low density lipoprotein. *Free Radic Res Commun*, 1989, 6(1), 67-75.

Expert Panel on Detection Evaluation, and Treatment of High Blood Cholesterol in Adults. Executive Summary of the Third Report of The National Cholesterol Education Program (NCEP) Expert Panel on Detection, Evaluation, And Treatment of High Blood Cholesterol In Adults (Adult Treatment Panel III). *JAMA*, 2001, 285(19), 2486-2497.

Fernandez, ML. Soluble fiber and nondigestible carbohydrate effects on plasma lipids and cardiovascular risk. *Curr Opin Lipidol*, 2001, 12(1), 35-40.

Friedewald, WT; Levy, RI; Fredrickson, DS. Estimation of the concentration of low-density lipoprotein cholesterol in plasma, without use of the preparative ultracentrifuge. *Clin Chem*, 1972, 18(6), 499-502.

Fuentes, F; Lopez-Miranda, J; Garcia, A; Perez-Martinez, P; Moreno, J; Cofan, M; et al. Basal plasma concentrations of plant sterols can predict LDL-C response to sitosterol in patients with familial hypercholesterolemia. *Eur J Clin Nutr*, 2008, 62(4), 495-501.

Geelen, A; Zock, PL; Vries, JHd; Katan, MB. Apolipoprotein E polymorphism and serum lipid response to plant sterols in humans. *Eur J Clin Invest*, 2002, 32(10), 738-742.

Gigleux, I; Jenkins, DJ; Kendall, CW; Marchie, A; Faulkner, DA; Wong, JM; et al. Comparison of a dietary portfolio diet of cholesterol-lowering foods and a statin on LDL particle size phenotype in hypercholesterolaemic participants. *Br J Nutr*, 2007, 98(6), 1229-1236.

Graaf, JD; Nolting, PRDS; Dam, MV; Belsey, EM; Kastelein, JJ; Pritchard, PH; et al. Consumption of tall oil-derived phytosterols in a chocolate matrix significantly decreases plasma total and low-density lipoprotein-cholesterol levels. *Br J Nutr*, 2002, 88(5), 479-488.

Grundy, SM; Cleeman, JI; Daniels, SR; Donato, KA; Eckel, RH; Franklin, BA; et al. Diagnosis and management of the metabolic syndrome: an American Heart Association/National Heart, Lung, and Blood Institute scientific statement: Executive Summary. *Crit Pathw Cardiol*, 2005, 4(4), 198-203.

Gylling, H; Hallikainen, M; Kolehmainen, M; Toppinen, L; Pihlajamaki, J; Mykkanen, H; et al. Cholesterol synthesis prevails over absorption in metabolic syndrome. *Transl Res*, 2007, 149(6), 310-316.

Gylling, H; Miettinen, TA. Baseline intestinal absorption and synthesis of cholesterol regulate its response to hypolipidaemic treatments in coronary patients. *Atherosclerosis*, 2002, 160(2), 477-481.

Hallikainen, M; Toppinen, L; Mykkanen, H; Agren, JJ; Laaksonen, DE; Miettinen, TA; et al. Interaction between cholesterol and glucose metabolism during dietary carbohydrate modification in subjects with the metabolic syndrome. *Am J Clin Nutr*, 2006, 84(6), 1385-1392.

Hallikainen, MA; Uusitupa, MI. Effects of 2 low-fat stanol ester-containing margarines on serum cholesterol concentrations as part of a low-fat diet in hypercholesterolemic subjects. *Am J Clin Nutr*, 1999, 69(3), 403-410.

Hansel, B; Nicolle, C; Lalanne, F; Tondu, F; Lassel, T; Donazzolo, Y; et al. Effect of low-fat, fermented milk enriched with plant sterols on serum lipid profile and oxidative stress in moderate hypercholesterolemia. *Am J Clin Nutr*, 2007, 86(3), 790-796.

Hayes, KC; Pronczuk, A; Wijendran, V; Beer, M. Free phytosterols effectively reduce plasma and liver cholesterol in gerbils fed cholesterol. *J Nutr*, 2002, 132(7), 1983-1988.

Howard, BV; Gidding, SS; Liu, K. Association of apolipoprotein E phenotype with plasma lipoproteins in African-American and white young adults. The CARDIA Study. Coronary Artery Risk Development in Young Adults. *Am J Epidemiol*, 1998, 148(9), 859-868.

Hunninghake, DB; Stein, EA; Dujovne, CA; Harris, WS; Feldman, EB; Miller, VT; et al. The efficacy of intensive dietary therapy alone or combined with lovastatin in outpatients with hypercholesterolemia. *N Engl J Med*, 1993, 328(17), 1213-1219.

Ishiwata, K; Homma, Y; Ishikawa, T; Nakamura, H; Handa, S. Influence of apolipoprotein E phenotype on metabolism of lipids and apolipoproteins after plant stanol ester ingestion in Japanese subjects. *Nutrition*, 2002, 18(7-8), 561-565.

Jimenez-Escrig, A; Santos-Hidalgo, AB; Saura-Calixto, F. Common sources and estimated intake of plant sterols in the Spanish diet. *J Agric Food Chem*, 2006, 54(9), 3462-3471.

Jones, PJ. Inter-individual variability in response to plant sterol and stanol consumption. *J AOAC Int*, 2015 May 4, In press.

Jones, PJ; Raeini-Sarjaz, M; Ntanios, FY; Vanstone, CA; Feng, JY; Parsons, WE. Modulation of plasma lipid levels and cholesterol kinetics by phytosterol versus phytostanol esters. *J Lipid Res*, 2000, 41(5), 697-705.

Judd, JT; Baer, DJ; Chen, SC; Clevidence, BA; Muesing, RA; Kramer, M; et al. Plant sterol esters lower plasma lipids and most carotenoids in mildly hypercholesterolemic adults. *Lipids*, 2002, 37(1), 33-42.

Kassis, AN; Vanstone, CA; Abumweis, SS; Jones, PJ. Efficacy of plant sterols is not influenced by dietary cholesterol intake in hypercholesterolemic individuals. *Metabolism*, 2008, 57(3), 339-346.

Katan, MB; Grundy, SM; Jones, P; Law, M; Miettinen, T; Paoletti, R; et al. Efficacy and safety of plant stanols and sterols in the management of blood cholesterol levels. *Mayo Clin Proc*, 2003, 78(8), 965-978.

Kempen, HJ; Knijff, Pd; Boomsma, DI; Voort, HAvd; Leuven, JAG; Havekes, L. Plasma levels of lathosterol and phytosterols in relation to age, sex, anthropometric parameters, plasma lipids, and apolipoprotein E phenotype, in 160 Dutch families. *Metabolism*, 1991, 40(6), 604-611.

Knopp, RH; Retzlaff, B; Fish, B; Walden, C; Wallick, S; Anderson, M; et al. Effects of insulin resistance and obesity on lipoproteins and sensitivity to egg feeding. *Arterioscler Thromb Vasc Biol*, 2003, 23(8), 1437-1443.

Korpela, R; Seppo, L; Laakso, J; Lilja, J; Karjala, K; Lahteenmaki, T; et al. Dietary habits affect the susceptibility of low-density lipoprotein to oxidation. *Eur J Clin Nutr*, 1999, 53(10), 802-807.

Law, M. Plant sterol and stanol margarines and health. *BMJ*, 2000, 320(7238), 861-864.

Lefevre, M; Champagne, CM; Tulley, RT; Rood, JC; Most, MM. Individual variability in cardiovascular disease risk factor responses to low-fat and low-saturated-fat diets in men: body mass index, adiposity, and insulin resistance predict changes in LDL cholesterol. *Am J Clin Nutr*, 2005, 82(5), 957-63.

Lloyd-Jones, D; Adams, RJ; Brown, TM; Carnethon, M; Dai, S; Simone, GD; et al. Executive summary: heart disease and stroke statistics--2010 update: a report from the American Heart Association. *Circulation*, 2010, 121(7), 948-954.

Lottenberg, AM; Nunes, VS; Nakandakare, ER; Neves, M; Bernik, M; Santos, JE; et al. Food phytosterol ester efficiency on the plasma lipid reduction in moderate hypercholesterolemic subjects. *Arq Bras Cardiol*, 2002, 79(2), 139-142.

Mahley, RW. Apolipoprotein E: cholesterol transport protein with expanding role in cell biology. *Science*, 1988, 240(4852), 622-630.

Maki, KC; Davidson, MH; Umporowicz, DM; Schaefer, EJ; Dicklin, MR; Ingram, KA; et al. Lipid responses to plant-sterol-enriched reduced-fat spreads incorporated into a National Cholesterol Education Program Step I Diet. *Am J Clin Nutr*, 2001, 74(1), 33-43.

Mannarino, E; Pirro, M; Cortese, C; Lupattelli, G; Siepi, D; Mezzetti, A; et al. Effects of a phytosterol-enriched dairy product on lipids, sterols and 8-isoprostane in hypercholesterolemic patients: a multicenter Italian study. *Nutr Metab Cardiovasc Dis*, 2009, 19(2), 84-90.

Mataix, J; García, L; Mañas, M; Martinez-Victoria, E; Llopis, J. Tablas de composición de alimentos españoles (Spanish Food Composition Tables). 4th ed. Granada: *University of Granada*, 2003.

Matvienko, OA; Lewis, DS; Swanson, M; Arndt, B; Rainwater, DL; Stewart, J; et al. A single daily dose of soybean phytosterols in ground beef decreases serum total cholesterol and LDL cholesterol in young, mildly hypercholesterolemic men. *Am J Clin Nutr*, 2002, 76(1), 57-64.

Mensink, RP; Ebbing, S; Lindhout, M; Plat, J; Heugten, MMv. Effects of plant stanol esters supplied in low-fat yoghurt on serum lipids and lipoproteins, non-cholesterol sterols and fat soluble antioxidant concentrations. *Atherosclerosis*, 2002, 160(1), 205-213.

Miettinen, TA; Vanhanen, H. Dietary sitostanol related to absorption, synthesis and serum level of cholesterol in different apolipoprotein E phenotypes. *Atherosclerosis,* 1994, 105(2), 217-226.

Miller, NJ; Rice-Evans, C; Davies, MJ; Gopinathan, V; Milner, A. A novel method for measuring antioxidant capacity and its application to monitoring the antioxidant status in premature neonates. *Clin Sci (Lond)*,1993, 84(4), 407-412.

Miller, SA; Dykes, DD; Polesky, HF. A simple salting out procedure for extracting DNA from human nucleated cells. *Nucleic Acids Res*, 1988, 16(3), 1215.

Mussner, MJ; Parhofer, KG; von Bergmann, K; Schwandt, P; Broedl, U; Otto, C. Effects of phytosterol ester-enriched margarine on plasma lipoproteins in mild to moderate hypercholesterolemia are related to basal cholesterol and fat intake. *Metabolism*, 2002, 51(2), 189-194.

Naumann, E; Plat, J; Kester, AD; Mensink, RP. The baseline serum lipoprotein profile is related to plant stanol induced changes in serum lipoprotein cholesterol and triacylglycerol concentrations. *J Am Coll Nutr*, 2008, 27(1), 117-126.

Naumann, E; Plat, J; Mensink, RP. Changes in serum concentrations of noncholesterol sterols and lipoproteins in healthy subjects do not depend on the ratio of plant sterols to stanols in the diet. *J Nutr*, 2003, 133(9), 2741-2747.

Nichols, AV; Krauss, RM; Musliner, TA. Nondenaturing polyacrylamide gradient gel electrophoresis. *Meth Enzymol*, 1986, 128, 417-431.

Nigon, F; Lesnik, P; Rouis, M; Chapman, MJ. Discrete subspecies of human low density lipoproteins are heterogeneous in their interaction with the cellular LDL receptor. *J Lipid Res*, 1991, 32(11), 1741-1753.

Nissinen, MJ; Gylling, H; Miettinen, TA. Effects of dietary cholesterol and fat on serum non-cholesterol sterols according to different apolipoprotein E subgroups among healthy men. *Br J Nutr*, 2008, 100(2), 373-379.

Noakes, M; Clifton, P; Ntanios, F; Shrapnel, W; Record, I; McInerney, J. An increase in dietary carotenoids when consuming plant sterols or stanols is effective in maintaining plasma carotenoid concentrations. *Am J Clin Nutr*, 2002, 75(1), 79-86.

Noakes, M; Clifton, PM; Doornbos, AM; Trautwein, EA. Plant sterol ester-enriched milk and yoghurt effectively reduce serum cholesterol in modestly hypercholesterolemic subjects. *Eur J Nutr*, 2005, 44(4), 214-222.

Ntanios, FY; Duchateau, GS. A healthy diet rich in carotenoids is effective in maintaining normal blood carotenoid levels during the daily use of plant sterol-enriched spreads. *Int J Vitam Nutr Res*, 2002, 72(1), 32-39.

Olmedilla, B; Granado, F; Rojas-Hidalgo, E; Blanco, I. A rapid separation of ten carotenoids, three retinoids, alpha-tocopherol and d-alpha-tocopherol acetate by high performance liquid chromatography and its application to serum and vegetable samples. *J Liq Chromatogr*, 1990, 13(8),1455-1483.

Ooi, EM; Ng, TW; Chan, DC; Watts, GF. Plasma markers of cholesterol homeostasis in metabolic syndrome subjects with or without type-2 diabetes. *Diabetes Res Clin Pract*, 2009, 85(3), 310-316.

Ooi, EM; Watts, GF; Barrett, PH; Chan, DC; Clifton, PM; Ji, J; et al. Dietary plant sterols supplementation does not alter lipoprotein kinetics in men with the metabolic syndrome. *Asia Pac J Clin Nutr*, 2007, 16(4), 624-631.

Ordovas, JM. Genetic influences on blood lipids and cardiovascular disease risk: tools for primary prevention. *Am J Clin Nutr*, 2009, 89(5), 1509S-1517S.

Ostlund, RE; Jr. Phytosterols, cholesterol absorption and healthy diets. *Lipids*, 2007, 42(1), 41-45.

Packard, CJ. Small dense low-density lipoprotein and its role as an independent predictor of cardiovascular disease. *Curr Opin Lipidol*, 2006, 17(4), 412-417.

Pihlajamaki, J; Gylling, H; Miettinen, TA; Laakso, M. Insulin resistance is associated with increased cholesterol synthesis and decreased cholesterol absorption in normoglycemic men. *J Lipid Res*, 2004, 45(3), 507-512.

Plat, J; Brufau, G; Dallinga-Thie, GM; Dasselaar, M; Mensink, RP. A plant stanol yogurt drink alone or combined with a low-dose statin lowers serum triacylglycerol and non-HDL cholesterol in metabolic syndrome patients. *J Nutr*, 2009, 139(6), 1143-1149.

Plat, J; Mensink, RP. Relationship of genetic variation in genes encoding apolipoprotein A-IV, scavenger receptor BI, HMG-CoA reductase, CETP and apolipoprotein E with cholesterol metabolism and the response to plant stanol ester consumption. *Eur J Clin Invest*, 2002, 32(4), 242-250.

Plat, J; Onselen, ENv; Heugten, MMv; Mensink, RP. Effects on serum lipids, lipoproteins and fat soluble antioxidant concentrations of consumption frequency of margarines and shortenings enriched with plant stanol esters. *Eur J Clin Nutr*, 2000, 54(9), 671-677.

Polagruto, JA; Wang-Polagruto, JF; Braun, MM; Lee, L; Kwik-Uribe, C; Keen, CL. Cocoa flavanol-enriched snack bars containing phytosterols effectively lower total and low-density lipoprotein cholesterol levels. *J Am Diet Assoc*, 2006, 106(11), 1804-1813.

Quilez, J; Rafecas, M; Brufau, G; Garcia-Lorda, P; Megias, I; Bullo, M; et al. Bakery products enriched with phytosterol esters, alpha-tocopherol and beta-carotene decrease plasma LDL-cholesterol and maintain plasma beta-carotene concentrations in normocholesterolemic men and women. *J Nutr*, 2003, 133(10), 3103-3109.

Ras, RT; Hiemstra, H; Lin, Y; Vermeer, MA; Duchateau, GS; Trautwein, EA. Consumption of plant sterol-enriched foods and effects on plasma plant sterol concentrations--a meta-analysis of randomized controlled studies. *Atherosclerosis*, 2013, 230(2), 336-346.

Ridker, PM; Rifai, N; Cook, NR; Bradwin, G; Buring, JE. Non-HDL cholesterol, apolipoproteins A-I and B100, standard lipid measures, lipid ratios, and CRP as risk factors for cardiovascular disease in women. *J Am Med Assoc*, 2005, 294(3), 326-333.

Sanchez-Muniz, FJ; Maki, KC; Schaefer, EJ; Ordovas, JM. Serum lipid and antioxidant responses in hypercholesterolemic men and women receiving plant sterol esters vary by apolipoprotein E genotype. *J Nutr*, 2009, 139(1), 13-19.

Schaefer, EJ; Lamon-Fava, S; Johnson, S; Ordovas, JM; Schaefer, MM; Castelli, WP; et al. Effects of gender and menopausal status on the association of apolipoprotein E phenotype with plasma lipoprotein levels. Results from the Framingham Offspring Study. *Arterioscler Thromb*, 1994, 14(7), 1105-1113.

Schwab, US; Vogel, S; Lammi-Keefe, CJ; Ordovas, JM; Schaefer, EJ; Li, Z; et al. Varying dietary fat type of reduced-fat diets has little effect on the susceptibility of LDL to oxidative modification in moderately hypercholesterolemic subjects. *J Nutr*, 1998, 128(10), 1703-1709.

Seppo, L; Jauhiainen, T; Nevala, R; Poussa, T; Korpela, R. Plant stanol esters in low-fat milk products lower serum total and LDL cholesterol. *Eur J Nutr*, 2007, 46(2), 111-117.

Shrestha, S; Freake, HC; McGrane, MM; Volek, JS; Fernandez, ML. A combination of psyllium and plant sterols alters lipoprotein metabolism in hypercholesterolemic subjects by modifying the intravascular processing of lipoproteins and increasing LDL uptake. *J Nutr*, 2007, 137(5), 1165-1170.

Simonen, P; Gylling, H; Howard, AN; Miettinen, TA. Introducing a new component of the metabolic syndrome: low cholesterol absorption. *Am J Clin Nutr*, 2000, 72(1), 82-88.

St-Onge, MP; Lamarche, B; Mauger, JF; Jones, PJ. Consumption of a functional oil rich in phytosterols and medium-chain triglyceride oil improves plasma lipid profiles in men. *J Nutr*, 2003, 133(6), 1815-1820.

Sudhop, T; Sahin, Y; Lindenthal, B; Hahn, C; Luers, C; Berthold, HK; et al. Comparison of the hepatic clearances of campesterol, sitosterol, and cholesterol in healthy subjects suggests that efflux transporters controlling intestinal sterol absorption also regulate biliary secretion. *Gut*, 2002, 51(6), 860-863.

Sugiuchi, H; Uji, Y; Okabe, H; Irie, T; Uekama, K; Kayahara, N; et al. Direct measurement of high-density lipoprotein cholesterol in serum with polyethylene glycol-modified enzymes and sulfated alpha-cyclodextrin. *Clin Chem*, 1995, 41(5), 717-723.

Tammi, A; Ronnemaa, T; Miettinen, TA; Gylling, H; Rask-Nissila, L; Viikari, J; et al. Effects of gender, apolipoprotein E phenotype and cholesterol-lowering by plant stanol esters in children: the STRIP study. Special Turku Coronary Risk Factor Intervention Project. *Acta Paediatr*, 2002, 91(11), 1155-1162.

Thomsen, AB; Hansen, HB; Christiansen, C; Green, H; Berger, A. Effect of free plant sterols in low-fat milk on serum lipid profile in hypercholesterolemic subjects. *Eur J Clin Nutr*, 2004, 58(6), 860-870.

Thuluva, SC; Igel, M; Giesa, U; Lutjohann, D; Sudhop, T; von Bergmann, K. Ratio of lathosterol to campesterol in serum predicts the cholesterol-lowering effect of sitostanol-supplemented margarine. *Int J Clin Pharmacol Ther*, 2005, 43(7), 305-310.

Trautwein, EA; Duchateau, GSMJE; Lin, Y; Mel'nikov, SM; Molhuizen, HOF; Ntanios, FY. Proposed mechanisms of cholesterol-lowering action of plant sterols. *Eur J Lipid Sci Tech*, 2003, 105(3-4), 171-185.

Vanhanen, HT; Blomqvist, S; Ehnholm, C; Hyvonen, M; Jauhiainen, M; Torstila, I; et al. Serum cholesterol, cholesterol precursors, and plant sterols in hypercholesterolemic subjects with different apoE phenotypes during dietary sitostanol ester treatment. *J Lipid Res*, 1993, 34(9), 1535-1544.

Vanstone, CA; Raeini-Sarjaz, M; Parsons, WE; Jones, PJ. Unesterified plant sterols and stanols lower LDL-cholesterol concentrations equivalently in hypercholesterolemic persons. *Am J Clin Nutr*, 2002, 76(6), 1272-1278.

Varady, KA; St-Pierre, AC; Lamarche, B; Jones, PJ. Effect of plant sterols and endurance training on LDL particle size and distribution in previously sedentary hypercholesterolemic adults. *Eur J Clin Nutr*, 2005, 59(4), 518-525.

Victor, VM; Rocha, M. Targeting antioxidants to mitochondria: a potential new therapeutic strategy for cardiovascular diseases. *Curr Pharm Des*, 2007, 13(8), 845-863.

Von Bergmann, K; Sudhop, T; Lutjohann, D. Cholesterol and plant sterol absorption: recent insights. *Am J Cardiol*, 2005, 96(1A), 10D-14D.

Weststrate, JA; Meijer, GW. Plant sterol-enriched margarines and reduction of plasma total- and LDL-cholesterol concentrations in normocholesterolaemic and mildly hypercholesterolaemic subjects. *Eur J Clin Nutr*,1998, 52(5), 334-343.

Wolthers, BG; Walrecht, HT; Molen, JCvd; Nagel, GT; Doormaal, JJV; Wijnandts, PN. Use of determinations of 7-lathosterol (5 alpha-cholest-7-en-3 beta-ol) and other cholesterol precursors in serum in the study and treatment of disturbances of sterol metabolism, particularly cerebrotendinous xanthomatosis. *J Lipid Res*, 1991, 32(4), 603-612.

Wu, T; Fu, J; Yang, Y; Zhang, L; Han, J. The effects of phytosterols/stanols on blood lipid profiles: a systematic review with meta-analysis. *Asia Pac J Clin Nutr*, 2009, 18(2), 179-186.

Yagi, K. Simple assay for the level of total lipid peroxides in serum or plasma. *Methods Mol Biol,* 1998, 108, 101-106.

Yonekura, L; Nagao, A. Intestinal absorption of dietary carotenoids. *Mol Nutr Food Res*, 2007, 51(1), 107-115.

In: Phytosterols
Editor: Deanna Garner

ISBN: 978-1-63483-477-3


*Chapter 9*

# PHYTOSTEROLS DISTRIBUTION IN ALIMENTARY OIL AND FINGERPRINT

***Kelly S. Bezerra* and Nelson R. Antoniosi Filho***
Laboratório de Métodos de Extração e Separação (LAMES),
Instituto de Química, Universidade Federal de Goiás,
Campus Samambaia, Goiânia, GO, Brazil

## ABSTRACT

Phytosterols are lipid compounds present in eukaryotic cells of plant species. The intracellular synthesis of phytosterols part of a complex of reactions starting from terpenoid alcohols, thus forming the structure of squalene. The group of tetracyclic fused rings has a flexible side chain whose structural changes between sterols involve differences in this chain. The phytosterols include several different combinations of atoms resulting in different possible structures, which they were identified in several plant oils. The β-sitosterol is the main phytosterols, and we can not underestimating occurrence of stigmasterol and campesterol. Phytosterols may appear to genuinely and mainly as free and conjugated steroids (FS, ES, SG and ASG). However, the profile of steroids, or the types of steroids and their relative proportions are not equal, thus making their fingerprint of identification. These compounds have been highlighted because of the chemical and pharmaceutical applications, especially with medical-therapeutic purposes. Another function of phytosterols, and not least important, is their composition used in the characterization of oils and the fingerprint of the same. The characterization of oils by phytosterols analysis has been studied in order to detect the adulteration of noble oils. Various methods have been used in the isolation of steroids and subsequent identification and quantification of the same.

**Keywords:** oils, phytosterols, fingerprint

---

* Tel/FAX: + 55 62 3521-1221, Corresponding author: kiqarrezeb@hotmail.com.

# 1. INTRODUCTION

This chapter has a focus primarily on the description of phytosterols, how they are distributed in the oils and simple forms of identification, characterization and quantification of these constituents. The phytosterols lipid compounds are extracted from oleaginous plant species, the most abundant being, from over 250 different phytosterols, the β-sitosterol, stigmasterol, campesterol, brassicasterol and Δ5-avenasterol [1-2].

The total phytosterols are distributed in the range of 1000 to 5000 ppm depending on the source of plant species [3]. When oil seeds pass through the oil extraction process and subsequent chemical treatment of this oil, phytosterols continue present in your composition [4]. These phytosterols can be presented in different vegetable oils in different forms, and may be linked to glucose molecules and composition fatty acids present in the oils. The steroid molecules bound to other constituents of the medium are known as conjugated steroids. Once the free steroids, this is, without bound molecules are the most abundant in oils [5].

Phytosterols have been researched, and have gained importance due to their medical and therapeutic benefits, in that several publications have reported. Although phytosterols are widely distributed in plant species, many of them were and are constantly synthesized in the laboratory, for commercial purposes, such as marking foods with these constituents, for example [6].

An important typical of steroids is that the distribution is specific to each source of sterol, this is, the types of steroils and their relative proportions are not the same for different oils, becoming the fingerprint for the identification thereof [7]. The steroids identified in soybean oil and quantified the percentage of its content will not be the same for the olive oil or corn, e.g., [8]. From here comes the importance of identifying and quantifying constituents. This chapter mentions some of the most important methods for extraction, separation and analysis of steroids, applied to vegetable oils. The characterization of oils by phytosterols analysis has already been studied in order to detect the adulteration of noble oils [9]. The studies in the literature that determine the steroid content in different samples as fingerprint, make the determination of the amount of total steroids, considering all the ways that these constituents are distributed in oils, for the precise determination of conjugated steroids is more expensive. Furthermore, the content of conjugates compared to the free steroid is much lower and almost negligible [10]. As the standard of steroids is exclusive of each oil, its analysis allows identify, characterize and quantify the phytosterols content in a sample, and in terms of functions and health benefits, determine how a food source can contribute in terms of benefits as the its contents on steroids.

# 2. DEFINITION, DISTRIBUTION AND FUNCTIONS OF PHYTOSTEROIDS

Steroids are organic lipid molecules whose structure is based on the tetracyclic ring system and unsaponifiable alcohols. The molecule forms a rigid structure of four rings with a flexible side chain. In accordance with the recommendations of IUPAC 1989, the sterol molecules consist of four rings marked A, B, C and D (Figure 1a) with the numbering of the carbons (Figure 1b). This set of rings is usually attached to two methyl groups and a side chain up to seventeen carbon atoms. The steroid nucleus is essentially a planar structure with substituents either above or below the plane of the ring named (β) or (α) respectively [3, 11-12].

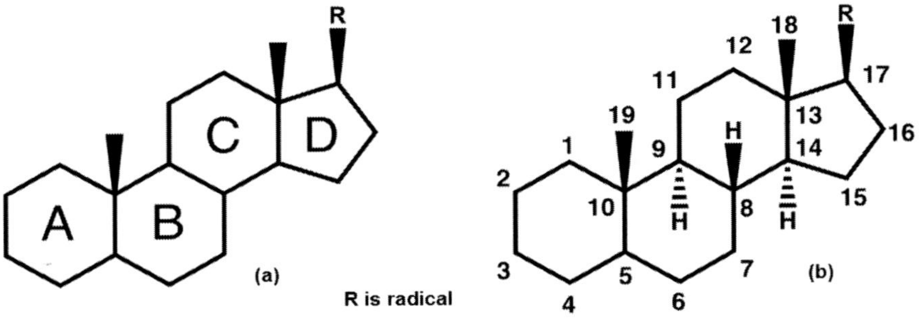


Figure 1. Representation of the set of rings that constitute the basic structure of steroids.

The stereochemistry of steroids depends on the conformation of the four rings system namely A, B, C, D. Rings A and B define the classification of steroids as α- or β- with fusion in *cis* and *trans* configuration in equal frequency. As for rings B and C, they are always in the *trans* conformation whereas C and D rings are only commonly in the *trans* conformation. Knowledge of the conformation of steroids is important especially for the knowledge of their physicochemical properties. In steroid 5α, the A-B rings are in *trans* configuration and the substituent is perpendicular to the plane of the ring, while in 5β steroids, the A-B rings are in *trans* configuration and the substituent is angled relative to the plane of the rings [12-14].

A free hydroxy sterol is an important factor that enables interactions with specific proteins in membrane phospholipids and [15]. For those containing one or more hydroxyl groups (OH) in the C-3 carbon and no carbonyl (C=O), the sterol nomenclature is given, therefore, almost all steroids have a 3β-hydroxyl group at C-3 position. The phytosteroids are largely hydrophobic having a polar group e.g., hydroxyl group on the third carbon atom, rendering them amphiphilic. According to the presence of a free hydroxyl group in phytosterols, they are appoint free phytosterols and are the principal and product of the phytosterol biosynthesis pathway [16-17].

The numbering of the core shown starts at the ring A, continues in B and C then ends in ring D (Figure 1). The numbers are used to describe positions in which the substituents and/or multiple bonds attached to individual compounds and due to the various possibilities of different constituents; different steroid compounds may be formed with structural similarity [7, 13]. The scientific names of steroids are given according to the number of carbon atoms in the side chain C-17, the number and position of the double bond in the ring system and the side chain [18-19].

Steroids are widely distributed in all eukaryotic organisms. Small variations in the molecular structure of steroids results in large differences in their effects as only very small amounts of these lipids exhibit considerable biological activity [20-22]. The most common and popularly known steroid is cholesterol that is present in animal cells, which is the metabolic precursor of steroid hormones such as mineral-ocorticoids, gluco-corticoids and bile acids [23-24]. As for the steroid ergosterol, it is synthesized by fungal cells, along with some other unicellular algae [25-26]. In contrast to animal and fungal cells, plant cells synthesize a complex matrix of mixtures of steroids, which have their own nomenclature, phytosteroids [27-29].

The phytosteroids are functional with respect to the growth of the plant, operating at the cellular level and regulating the water balance. They perform essential functions, regulating the fluidity and permeability of cell membranes in a similar manner to cholesterol in the membranes of animal cells [30-32]. The phytosterols have been isolated from a large number of species and according to many publications they probably exist in all species of gymnosperms and angiosperms [33-34]. Thus, a variety of phytosterols has been identified and studied, and are predominantly present in the oil plants [35-36].

The most abundant phytosterols are β-sitosterol, campesterol and stigmasterol. While β-sitosterol and stigmasterol are involved in membrane function, campesterol is the precursor for brassinosteroids, which are steroid plant hormones, needed for cell division and cell expansion [37-38]. It was related still that β-sitosterol and campesterol are more efficient in membrane permeability and the regulation of fluid and there is evidences that stigmasterol associate an important role in cell proliferation [3, 11]. Finally, it is noted that they influence the plant development and function through the location of the key-regulatory proteins [33].

The phytosteroids have both structural and metabolic functions where they associate a leading role in stabilizing the lipid bilayer of the membrane owing to the fact that they are integral membrane components and determine the characteristics of the plasma membrane and endoplasmic reticulum. Studies also show that, very likely, they have a certain function in the membrane adaptation to temperature variations [3, 6, 11]. In small doses, they are capable of producing a positive effect on the enzymatic transport and signaling processes in cells [1].

Unlike animals and fungi producing the tetracyclic lanosterol, plants produce the pentacyclic cycloartenol as the initial precursor steroids [5]. The sterols together with tocopherols, carotenoids and chlorophylls are formed by polymerization of isoprene units [39, 22]. They are family members of triterpene alcohols. Terpenoids are formed by cyclic structures that result from transformation of isoprenoid chains giving rise to them and the triterpenes constitute the largest and most important group of terpenoids [4, 15].

A route that starts with the reduction of HMG-CoA-reductase to mevalonate synthesizes all the triterpenes. Molecules are added to mevalonate diphosphate farnesyl, which are combined to assemble the structure of the squalene (30 carbons or "three-terpenes"). Enzymatic steps are carried out to ring closure, and then cycloartenol is formed (also 30 carbon atoms), resulting in the structure of brassinosteroids and phytoestroids. These transformations consist of a large number of reactions, such as reductions, oxidations, cyclizations (involving the formation of carbocations), ring breaks or rearrangements, which place an enormous structural diversity to these metabolites. Thus, steroids are synthesized intracellular from a complex series of reactions and with the involvement of intermediates [6, 12, 23, 28].

According to the structural base and number of methyl groups in the C-4 phytosteroids, they can be divided into three groups: two (4-dimethyl), one (4-monomethyl) or none (4-desmethyl). The most abundant are 4-desmethyl steroids, which are major structural components of plant cell membrane and include, for example, β-sitosterol, campesterol and stigmasterol.

Other phytosteroids as the 4-mono and 4,4-dimethyl sterols are synthesized during biosynthesis pathway, therefore they are steroid precursors and generally occur in relatively minor amounts. Most 4-demethyl phytosterols (for example, stigmasterol, β-sitosterol, campesterol, brassicasterol and avenasterol) have a double bond between carbons C-5 and C-6 of the ring system and are therefore termed as Δ5-phytosteroids. Yet another group of base demethyl-steroids which are abundant in certain oleaginous plants have a double bond between carbons 7 and 8 instead of 5 and 6, and therefore are referred Δ7 phytosteroids [6, 12, 17, 25].

In addition to their broad structural changes, resulting from substitutions on the side chain and the number and position of double bonds in the tetracyclic rings, different phytosterols have various roles in higher plants. They include various combinations of atoms forming different types of steroids. The fusion of the rings from squalene still allows the formation of a large number of possible isomers [11, 23].

Phytosterols having a 3β-hydroxyl group in its free structure is termed as free phytosterol (FS), in other words, it is not derivative whereas conjugated sterols have their 3β-hydroxyl group linked covalently with other constituents (Figure 2). The phytosterols may occur as esters of sterols (ES), with the 3β-hydroxyl group esterified with a long chain saturated or unsaturated fatty acids, mainly oleic and linoleic acids at, or phenolic acids. Sterol glycosides (SG) are formed when the 3β-hydroxyl group is attached to a monosaccharide, usually glucose. When that monosaccharide is at the C-6 position and esterified with a fatty acid, it forms an acylated steroid glycosides (ASG). These are the main forms of distribution of phytosterols [35, 40-41].

All steroids are completely soluble in oily chemistry composition. Sterols present in the form of SG and ASG are more polar than those present in the ES and S forms, mainly due to glycosidic chain present in their structures. Metabolic studies show that inter-conversion of free sterols and their conjugates is very fast [41].

Recently, have reported that β-sitosterol glucoside works as a primer for cellulose synthesis in plants [12]. In general, steroid in the form of free or conjugated steroids, are distributed in the range of 1000 to 5000 ppm in oil [11].

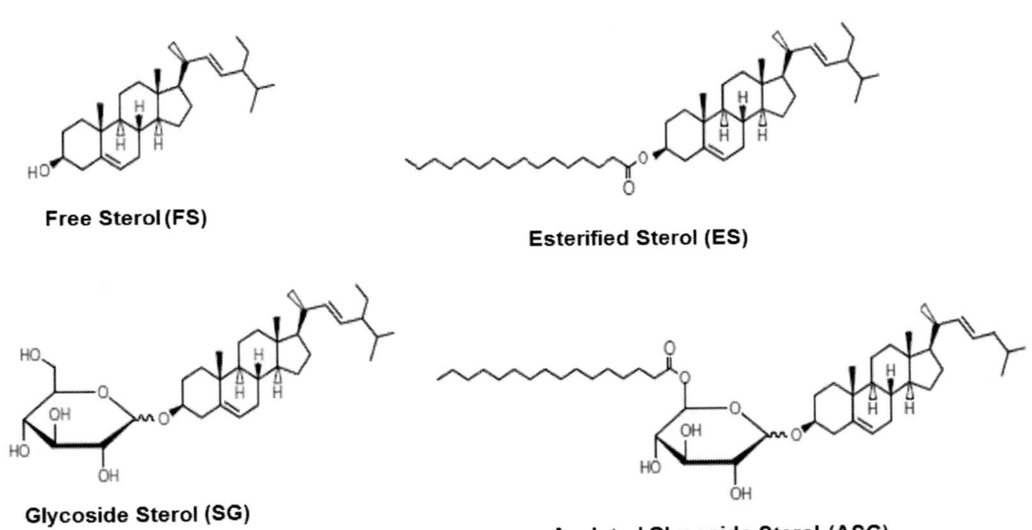


Figure 2. Representation of the main forms of steroids in vegetable oils: free sterol (FS), esters of sterol (ES), sterol glycoside (SG) and acylated sterol glycoside (ASG).

# 3. Nutricional Interest of Phytosteroids Content in Vegetable Oils

For years, many researchers have concentrated atention in the study of cholesterol levels, since their high levels prove to be an important risk factor for cardiovascular diseases [42]. To minimize and/or reduce cholesterol levels in the blood, are used dietary restrictions associated with the use of drugs which inhibit biosynthesis of cholesterol levels in humans [18, 27, 42]. The prospect of lowering cholesterol levels, consuming functional foods enriched with phytonutrients looks more attractive than the use of drugs or dietary restrictions. Has been considerable interest in the marketing of products derived from phytoesteroids for this purpose [42-43]. Diets rich in β-sitosterol, for instance, has been used since then to reducing cholesterol levels in individuals with hypercholesterolemia [44].

Due to the poor solubility and bioavailability of the free phytosterols, the effect of lowering of cholesterol levels in serum are not always consistent and very high doses (up to 25-50 g $day^{-1}$) are at times required for efficacy. This problem of solubility and bioavailability has made the use of nutritional phytosterols rapidly decreased [12, 38, 44].

The fact is that research has progressed and currently issues regarding solubility and bioavailability of phytosterols started to decrease and be subject to control in nutritional studies. With the advent research into the importance of phytosterols as food components, many food companies are actively developing enriched foods with phytosterols [45]. Phytosteroids derived from oils currently represent a stock of important basis for the health and nutrition industry [46].

The phytosterols have been the subject of interest not only to hypocholestenemic effect but also for the anti-inflammatory, antioxidant, anti-carcinogenic and immunomodulation [47-49]. Phytosterols are also used in the treatment of dermatitis and other diseases by stimulating the healing process [37]. Its effectiveness in reducing the intestinal absorption of blood cholesterol and evince protection against cardiovascular disease have been showed as the most extensively reported [44, 48-49].

When investigating the accumulation of phytosteroids embedded in the cell membranes of many tissues, it was found that phytosteroids tend to accumulate in adipose tissue in hepatic microsomes, as well as promoting a reduction of 25% of low low density lipoprotein (LDL) in the tissues [48]. Epidemiological studies suggest that phytosteroids consumed in a diet may provide protection against the most common types of cancer, such as colon, breast and prostate cancer [47-48].

Currently, there is a need for further studies to elucidate the mechanisms by which phytosteroids influence in cell growth and apoptosis of tumor cells, and to identify the concentration of phytosteroids capable of promoting cancer protection is the same capable to promote hypocholestenemic, for example [44, 49].

Among the different phytosteroids, β-sitosterol was the more intensively investigated with regard to its physiological effects in humans by be present at high levels and almost all plant-derived foods and oils [39]. Thus, mix of oils composed of high concentration of phytosteroids can confer nutritional benefits and efforts have been attempts by the food industry to develop functional food enriched with phytosterols [31, 44]. All these nutritional points addressed has led to interest in quantifying phytosteroids in different foods such as in oils [22].

# 4. Distribution Content of Steroids in Various Vegetable Oils – Fingerprint

Some methods based on analysis of fatty acid composition are used in the characterization of various vegetable oils. However, the analysis of composition of fatty acids or triacylglycerides can be inefficient in the characterization of an oil, since most oils may contain the same fatty acids or triacylglycerides [50]. Nevertheless, analysis of the composition of other components can be useful to identifying the various sources of vegetable oils. One group of chemicals that can be used for this proposal are phytosterols, which are present along with other groups of compounds forming the unsaponifiable matter. The unsaponifiable matter consists of different chemical components, and is present in all oils, which industrial processes for the conversion of oil seed into oil are not able to completely eliminate these constituents [23, 52].

The vegetable oils are used as a source for extraction of phytosteroids. High content of phytosteroids are found in seeds of oil plants. Depending on the oil source, a large variation in content and distribution of steroids can be seen, but its composition varies in a narrower range compared to triacylglycerides and of fatty acid esters, for example [51]. The called steroids profile, this is, the types of steroids and their relative proportions are not equal in different oils and fats, thus becoming a fingerprint for identification.

As the composition of fatty acids in vegetable oils cannot distinguish between a different kind of oil or different qualities of a given type oil, the phytosterols fraction can be analyzed and used to detect the addition of a declared cheap oil for more expensive oils (e.g., canola oil in olive oil) or for distinguishing between different qualities of the same oil (e.g., from virgin, i.e., cold pressed, and the extraction of olive oils or oils subjected to different treatments) [6, 11, 28]. Another advantage of phytosteroids profiles is that a small sample volume may be used for simultaneous measurement of a large number of phytosteroids, whereas the test for individual steroid separately generally requires a larger sample volume [26, 33]. β-sitosterol, campesterol and stigmasterol are predominant steroids while others phytosteroids are usually present in only certain families of plants or even species. Brassicasterol, for example, it is present only in seeds of the *Brassicaceae* family, and therefore can be used to identify this family with the original oils from other oil seeds [12].

The level of purity an oil not significantly affects the relative proportions of steroids but affects the total content of these steroids. This can be used as a means to identifying the oil thus to detect tampering. The studied by the composition of steroids the adulteration of olive oil - considered a noble oil and substantially high value - through other cheaper vegetable oils such as cottonseed oil, sunflower, corn and soybeans. As a steroids pattern are exclusive of each oil, its analysis is to promote the quality control of these oils [35, 52].

Research shows that quantitative analysis of steroid conjugates did not change considerably in an oil suffering any processing or treatment, unlike the free steroids [28, 37]. The steps of the refining process as degumming, neutralization, bleaching and deodorizing can lead to a significant reduction in the content of free sterols present in these oils [26, 39]. Several factors may contribute to the loss of phytosteroids, including adsorption, partitioning, oxidation and dehydration [53].

The composition of total phytosterols is calculated from the free sterols, esterified and their glycosylated forms. Extracted oils contain considerably higher concentrations of free steroids (FS) than cold pressed oils, whereas the esterified steroid (ES) are almost completely recovered when the oil is cold pressed [35]. Treatments such as filtration or crystallization, affect the free and esterified steroids to different extents [23]. The degumming step with water, for example, can effectively remove steroidal glycosides (SG and ASG) to a non-detectable level [40]. During the deodorization process, a significant reduction in the total content of phytosterols was reported, mainly due to distillation [29]. At the same time, has been observed an increase in the content of esterified steroids (ES) after the deodorisation process. It is important to note that in the literature the influence of the refining process of vegetable oil and how it affects the phytosterols is still both scarce and incomplete. The papers published about the influence of refining in phytosterols composition in vegetable oils, show that in general this reduction occurs between 40% - 60% [25].

The entire contents of phytosteroids may also be affected by genetic modification. They compared ten transgenic canola genotypes and non-transgenic experimental and the content in phytosteroids was influenced by the genetic modification [33]. In summary, there was good agreement between the analysis of content of free and esterified phytosteroids (ES and FS) compared with the content of total phytosteroids by analysis of more than ninety samples [35]. During seed germination, the phytosterols content increases, which is due to intensive in membrane biosynthesis and phytosterols that accumulate in seeds and meristematic tissue and participate of important role in cell proliferation and differentiation [46]. Consequently, the phytosterols synthesis appears mostly during seed formation and sprouting, and its contents will provide a supply to the growth of new cells and young shoots. At the moment the seeds are ripe or the tissues older, the intensity of the biosynthesis decreases [24].

The high total phytosterols content of a respective cultivation appears not to be related to their high oil content, because apparently no correlation between oil content and the total content of phytosterols [16]. High expectations have been presented with respect to phytosterols analysis and breeding applications in developing traditional plant cultivars with phytosterols desirable composition [33].

The composition of phytosterols is currently the analytical method of choice for determining the species of plant that the oil is extracted [25]. The studied the variation in lower components of free oil esterified with maturation and storage at room temperature. They found that the total content of steroids varied more in oils stored for different periods than those obtained at different stages of maturation [23-34]. The results suggest that the influence of storage temperature on phytosterols composition is more important than the influence of storage time [27]. To test the effect of temperature between flowerings to maturation of seeds on the phytosterols accumulation of seeds of each of the nine soybean lines were spread and the plants that were maintained in an oven between 18 and 35°C.

After flowering, two plants of each line were placed in growth chambers. They used three daily temperature regimes: 35, 27 and 20°C for 12 h light-cycle alternating with 25, 17 and 10°C respectively for 12 h dark cycle [6, 35]. Thus, changes in membrane constituent in response to cold stress may have little relevance to regulation of phytosterols content in oil biosynthesis temperature. The total phytosterols levels are consistently higher than the high temperatures [21].

All these results shows that the variation in the content of phytosterols in vegetable oils may be susceptible to several variations from the growth of the plant, to the extraction of oil from seed, ending in the industrial process of improving the same. However it is noteworthy that despite the inconsistency of the quantities of total oil in these steroids, their relative proportions and their types also become peculiar to each vegetable oil.

# 5. Methods of Extraction, Separation and Analysis of Steroids In Oils

Currently there is great interest in the development of fast analysis methods of pytosteroids. Many attempts are described in the technical and scientific literature to determine steroids in their intact forms (S, ES, SG and ASG) [12, 45, 54]. Most studies analyze free steroids (S) despite that SG, ASG and ES forms are important in determining the composition of steroids in vegetable oils. Among the methods of isolation of steroids already known, the conventional method involves the saponification of the sample followed by extracting the unsaponifiable fraction of organic solvent and/or pre-fractionation techniques such as liquid chromatography column (LC) thin layer chromatography (TLC) and solid phase extraction (SPE) [12, 35, 45].

The most used methods for the separation of steroids are the chromatographic methods, they are gas chromatography (GC) and high performance liquid chromatography (HPLC). The detectors coupled to these techniques and most commonly used are flame ionization detector (FID) to mass spectrometry (MS), mass spectrometry tandem (MS/MS), Uv-vis, diodes array detector (DAD), among others [37, 55].

Apparently, there is no single sample preparation technique that can be universally applied for steroids analysis. The selection of a type of sample preparation depends on the sensitivity required, the kind of matrix that these steroids are inserted and the required method performance. Thus, the development of rapid and reliable analytical methods for quality control, including the identification of several phytosteroids in oils are important and challenging.

## 5.1. Extraction of Phytosteroids by Liquid Cromatography (LC)

Traditionally, liquid chromatography (LC) is used to fractionate steroids. The extraction of free steroids and glycosides are performed with hexane, methylene chloride, ethanol, and isopropanol to SF, SE, SG and ASG, respectively. SG and ASG is partially extracted with hexane with a more polar solvent. Increasing the polarity of the extractor solvent the extraction percentage increase. Chloroform-methanol in varying ratios can also be used for the extraction of all lipids classes of steroids [12]. Methods involving column chromatography on silica or cleaning cartridges for removing different acylglycerides with solvent of different polarity are time consuming and impractical [57].

Some methods extracted the free and esterified phytosteroids are separated by liquid chromatography on silica column by elution with hexane/ethyl acetate, followed by hexane/diethyl ether/ethanol. Elution with hexane/ethyl acetate to collect the fraction of ES and then elution with hexane/diethyl ether/ethanol to collect the fraction SF [57].

In the second step, the silica column is eluted with a polar solvent, allowing the recovery of free sterols. The experiments are carried out with different mixtures of non-polar solvents.

The polarity of the nonpolar solvent, which consisted of hexane, is modified by addition of small amounts of ethyl acetate as the polar modifier. Free sterols are eluted in the second fraction by elution with a polar solvent mixture of hexane/diethyl ether/ethanol. The elution of the silica gel column with a ratio of hexane/ethyl acetate is found to give a good separation of all esterified sitosterol stearate in the first fraction and free steroid in the second fraction. These data indicate that it is possible to separate the free sterols and esterified with a minimal amount of solvent on a small silica column. However, a good polarity of the first solvent is great importance [58]. To achieve a good separation between saturated and steroids, it is necessary to optimize the volumes of the different fractions. For this reason, the volume of the first fraction should be adjusted so that when the second fraction is analyzed and the peaks representing the saturated hydrocarbons are of a lower intensity; if they do not appear, but the pattern of peak intensity is low, the volume should be reduced [59].

The main problem for the LC is the concern that the separated components are at maximum separation efficiency in order to obtain a strait fraction containing components of interest [57].

## 5.2. Identification of Phytosteroids by Thin-Layer Chromatography (TLC)

Preparative thin layer chromatography (TLC) can be used to separate and fractionate steroids [12]. TLC continues to be an important method for the qualitative analysis of steroids due to their inherent advantages - many samples can be analyzed simultaneously and quickly and at low cost, and various techniques for separation and detection procedures may be applied in addition to low detection limits. The TLC is a practical solution generally accepted to characterize extracts enriched with active constituents such as steroids [59].

Some methods extracted analyzed the sterols in the TLC plate eluted with heptane/ethyl acetate and the spots are visualized. The analytes are extracted with diethyl ether [60]. The zone corresponding to phytosterols are scraped from the plate and eluted with dichloromethane. Phytosterols are converted to their corresponding trimethylsilyl ether derivatives, by adding a derivatizing mixture comprised of pyridine/ hexamethyl-disilazane/trimethylchlorosilane for pre-analysis and heated at 60°C for 15 min, for exemple [30, 59].

The solubilized sample in chloroform can be subjected to thin layer chromatography on silica using acetic acid/acetone/water as developing solvent. The isolation of the ASG is separated and solubilized in silicic acid. The mixture is then diluted with water and extracted with diethyl ether. The isolated ASG can be separated and solubilized in silicic acid. The ASG purified sample can be treated with derivatization by benzoyl chloride in pyridine solution for 30 min [61].

## 5.3. Extraction of Phytosteroids by Solid Phase Extraction (SPE)

Sample preparation included a step of pre-extraction in solid phase. The pre-fractionation or purification of the sample by SPE resulted in improved ionization efficiency for all classes of steroids, in particular for the more nonpolar. The addition of this step resulted in increased sensitivity and reduced matrix effects during ionization.

Solid phase extraction (SPE) can be used to separate free and esterified steroids from glycoside steroids [62]. A silica SPE cartridge is conditioned with chloroform, and then the sample is transferred quantitatively to the cartridge. An additional of chloroform can be used to complete the elution of FS and SG is eluted with of methanol. The SG fraction is derivatized with pyridine and bis (trimethylsilyl) trifluoroacetamide. The fraction SG can be derivatized with pyridine and bis (trimethylsilyl) trifluoroacetamide [62]. The separation of free sterols and glycosides by SPE before saponification allows fraction glucosidic to be directly derivatized, thereby is reducing the potential for incomplete recovery during post saponification extraction [62].

Some methods use SiOH-SPE cartridge activated with hexane and the cartridge is washed with hexane/diethyl ether and hexane/diethyl ether to remove nonpolar compounds. The steroids are eluted with of acetone. Cholestanol or 19-OH-cholesterol are added to these acetone extracts to act as reference standards and the extracts are then drier and dissolved in pyridine and BSTFA/TMCS for derivatization [64].

For SPE initially, can be used C18 and SiOH cartridges for solid phase extraction. The C18 cartridge is activated with methanol and water. The sample thus eluted whit vacuum. The steroid fraction can be eluted with methanol and chloroform. The SiOH cartridge are activated with hexane. After filtration and sample application, the cartridge is rinsed with hexane and of a mixture of hexane and diethyl ether. The steroid fraction are eluted with of a mixture of hexane/diethyl ether. The samples are derivatized with BSTFA/ TMCS in pyridine [23].

## 5.4. Extraction of Phytosteroids by Acid and Alkaline Hydrolysis

Traditional methods of analysis by hydrolysis of steroid is used for converting the acylglycerides in saponified fatty acid. The fraction of interest is so separated from the other by extraction with organic solvent. These classical methods, however, can only determine the total sterols. The analytical methods involving basic hydrolysis (saponification) and acidic, eventually converts ES to ASG and FS to SG, both by rupture of the steroid ester bond, causing an important lost of information on the actual composition of steroids due to the hydrolysis process [12, 45].

SE may be hydrolyzed by saponification (alkaline hydrolysis with NaOH or KOH), while glycosidic bonds in SG and ASG require acidic hydrolysis (HCl). Once the vegetable oil samples generally contain SE and mainly little or no ESG or SG, the alkaline hydrolysis alone is sufficient to cleave all conjugates phytosteroids. Clearly, further investigation is necessary to improve the phytosterols hydrolysis methods. Enzymatic methods can be studied as a way to replace them because it has generated the best results compared to acidic and/or alkaline hydrolysis [12].

The saponification reaction is start by weighing of sample in a test tube and adding ethanol and potassium hydroxide (KOH) 70-50% (w/v) aqueous solution. The test tube is allowed to react in a previously sealed water bath at 100-60°C for 40-60 min. After the reaction time, the reaction product is wash successively with ethanol and water. The unsaponified matter is extract five times with portions of hexane or other non-polar organic solvent. These extracts are combined and washed with ethanol aqueous solution 10% (v/v).

The solution is drier with anhydrous sodium sulfate ($Na_2SO_4$) and the solvent is evaporate. The sample is solubilize in of pyridine and MSTFA (derivatizing agent) is added [64].

In oil is added the standard solution of cholesta-3,5-diene and alcohol 10% KCl keeping the mixture under reflux for 30 minutes. Add water and hexane. Stir the mixture vigorously for 30 seconds and allow the phases to separate. Collect the unsaponifiable phase and wash it with hexane and a mixture of ethanol/water until neutral pH is reached. Filter the solution through anhydrous sodium sulfate, washing with hexane and evaporate the solvent to dryness [31]. Is added to the sample together with 5α-cholestan-3-beta-ol as an internal standard. The unsaponifiable fraction was extracted three times with ether [22, 34].

The tubes are maintained in a water bath for 30 min at 70°C, and are cooled add water and hexane. The tubes are agitated vigorously and then centrifuged at 2,000 rpm for 10 min. The hexane layer is removed and the centrifugation treatment is repeated with addition of hexane. The hexane extracts are combined and the solvent are evaporated. Subsequently, the extract is dissolved in ethanol and analyzed [32].

The extraction of the unsaponifiable fraction can be performed by adding cholestanol to the sample as an internal standard with chloroform, and the mixture is subjected to solvent evaporation. The mixture is transferred to a separating funnel with distilled water and diethyl ether. The mixture is vigorously shaken and two phases are allowed to separate. The operation is repeated two more times with the extraction phase with ether. The ether fractions are combined in a funnel and washed with distilled water. The solution is dried with anhydrous sodium sulfate and filtered. The residue is allowed to cool, and then is weighed [27].

Potassium hydroxide (KOH) in ethanol can be added for saponification of the same sample, the mixture is maintained under ultrasound during for 10 min. The samples are incubated at 60°C for 1 h, then at ambient temperature for 12 h. Thereafter water is added and the unsaponifiable matter is extracted with ethyl ether. The ether extracts are combined and washed with water portions. The ether is evaporated and the sample is dissolved again in a bis (trimethylsilyl) trifluoroacetamide (BSTFA) as a derivatizing agent [33].

The sample preparation procedure containing conjugated steroids (ES, SG and ASG) included acidic and alkaline hydrolysis to release the steroid in the free form (FS) before analysis. Previously, plant sterol content obtained without the acid hydrolysis step is reported, as 64-80% of the samples are prepared with this treatment. For acid hydrolysis, 6 M HCl is added together with the sample into test tubes and placed in a shaking water bath at (80-85°C) for 60 min. After the sample is cooled, and lipids are extracted with hexane/diethyl ether maintaining the tube under stirring for 10 min after mixing. The organic layer is separated and the solvent is evaporated. For saponification, ethanol is added to the residue and the solution is transferred to a test tube. After addition of saturated aqueous KOH solution, the contents of the tubes are centrifuged and placed in a water bath with stirring (80-85°C) for 30 min and then cooled. For the extraction of the unsaponifiable fraction of water and cyclohexane are added, followed by centrifugation for 10 min. The sample is evaporated and dissolved in nitric acid and chloroform [39].

Hydrochloric acid (HCl) in methanol can be used to hydrolyze samples containing steroid ASG. The mixture is heated at 78°C for 1 h then water is added, followed by two extractions with pentane. The water is removed with anhydrous sodium sulfate and the combined extracts with sodium sulfate [53].

Some methods are developed on the basis of test tube microscale transposition of the conventional saponification reaction. This approach provides speed and ease of operation. Traditional methods, on the other hand, use reflux under heating, which requires more complex experimental apparatus than with a test tube and also require large quantities of sample and reagents [38].

## 5.5. Separation of Phytosteroids by Gas Chromatography (GC)

Gas chromatography (GC) has been widely applied due to its sensitivity and accuracy. However, the GC method has some disadvantages such as the need of derivatization of steroid [23]. The derivatization process involves replacing the active hydrogen of the hydroxyl groups of steroid. The derivatization provides a more nonpolar character, from alkylating reactions, acylation or silylation and increased volatility of these compounds, providing well defined chromatographic peaks and enhancements sensitivity and/or analytical specificity.

Chromatography method of quantification the free steroids is developed from the procedure which uses internal standard betulin. In quantitative analysis betulin solution is added to vial before derivatization. The analysis of free steroids is performed with a flame ionization detector (FID). A DB17HT capillary column (30 m X 250 μm X 0.15 μm) is used. The injector temperature is 280°C, the oven operated initially at 200°C and is heated to 330°C at a rate of 3°C $min^{-1}$. The detector operated at a temperature of 320°C [65].

Some studies, a rapid and simple method is developed for determining phytosteroids without derivatization. Gas chromatography is performed for this pupose by direct injection in a nonpolar megabore column DB-1 (0.53 x 5 mm x 1 μm). Fourteen samples, including olive oil, sunflower oil, peanut oil and soybean oil are quantitatively analyzed for the levels of total and free steroids. The injector and detector temperatures are 280°C and 360°C respectively. The initial oven temperature is set at 190°C for 3 min. The temperature is then set to 210°C at 1.5°C $min^{-1}$, maintained at 210°C for 2 min, increased to 230°C at 2°C $min^{-1}$, and finally increased to 300°C 30°C $min^{-1}$. [57].

GC separations can be performed using a CE -5 column (30m x 0.25 mm x 0.25 μm), using He as carrier gas at a pressure of 124.8 kPa and injection in split mode in a ratio of 15:1. The GC operating parameters are as follows: the initial oven temperature is 290°C. After an isothermal period of 20 minutes, the oven temperature is increased by 10°C $min^{-1}$ to 300°C and kept at this temperature for 10 min. Detection is performed by flame ionization with temperature detector set to 360°C [22].

A gas chromatographic analysis are performed using capillary column ZB-5 (30 x 0.25mm x 0.10 μm) Hydrogen is used as carrier gas at a pressure of 125 kPa. Injection in split mode is used at a temperature of 320°C. The oven temperature profile is: initial temperature 240°C and increased in 5°C $min^{-1}$ to final temperature of 265°C and maintained at this temperature for 10 min [2]. The GC analyzes are performed using a GC-FID, with PTV (Programmed Temperature Vaporization) injector and a HP-5 capillary column (15 x 0.25mm x 0.22 μm) hydrogen is used as carrier gas. The injector is operated at 350°C in the split injection mode. The oven temperature profile is: initial temperature of 75°C to 250°C at rate 40°C $min^{-1}$ and maintained at this final temperature for 30 min [30].

Some methods by GC analysis of samples performed using various steroids capillary columns: column coated with DRL (60 mx 0.25 mm x 0.12 μm) and DR5 column (30 m x 0.25 mm x 0.12 μm); Stabilwax column (60 m x 0.25 mm x 0.12 μm); DE225 (15 m x 0.25 mm x 0.12 μm); RTX-1701 (15 m x 0.25 mm 0.12 μm); CP-Wax 57 CB (25 m x 0.32 mm x 0.20 μm). The analysis is performed isothermally using split injection mode 1:50 with He as carrier gas linear velocity 40 cm $s^{-1}$. The temperature of the injector and flame ionization detector are maintained at 250°C and 290°C respectively. The comparison of the performance of polar and the nonpolar columns indicate significant differences. The nonpolar column showed no problems regarding resistance to high temperatures applied, unlike the polar columns. Also they showed better separations that the polar columns. The probability of co-elution is even greater with polar columns. These elution patterns, some of which have been noted earlier, indicate the need for monitoring the use of GC columns for the analysis of polar mixtures containing steroids [36].

GC-MS methods can be analyzed with interface is 270°C and an electronic impact ion source with 50-700 m/z at 70 eV. The temperature of the chromatographic oven is held isothermally at 250°C [36]. A RTX TG-65 column (60 mx 0.25 mm x 0.25 μm) is used. The temperatures of the injector and detector are maintained at 320°C. The oven temperature is programmed 200-300°C with a heating rate of 3°C $min^{-1}$ [37]. The samples are injected in splitless mode using a TM SAC-5 column (30 mx 0.25 mm x 0.25 μm). The initial oven temperature is 50°C for 2 minutes increasing to 250°C at a heating rate of 20°C $min^{-1}$, then from 250°C to 280°C at a rate of 1°C $min^{-1}$ and maintained for 3 min, and finally from 280°C to 305°C at 20°C $min^{-1}$ where it is maintained for 10 min. The detector is operated at 300°C. A cholestane is used as an internal standard [21]. The column used is Rtx-65TG (15 mx 0.25 mm x 0.10 μm) Helium is used as carrier gas at 125 kPa and a flow of 54 mL $min^{-1}$. The furnace temperature is held isothermally at 240°C and the temperatures of the injector and detector are maintained at 300°C. Splitless injection mode can be use [37].

## 5.6. Separation of Phytosteroids by High-Performance Liquid Chromatography (HPLC)

The high-performance liquid chromatography (HPLC) has advantages such as short analysis times and no need to derivatization of steroids, although some studies use this procedure [54]. In general, polar or nonpolar HPLC column ("normal/reverse phase") can be used to separate the phytosterols (for example, C18, ODS, phenyl, or C8). Although some methods have been reported of simultaneous analysis of both types of columns, nonpolar for FS analysis and SE, polar for SG and ASG analysis, experiments show that the most accurate way to quantify these lipids is to analyze these classes of polar steroids and apolar separately [39, 66-67].

Several high performance liquid chromatographic (HPLC) methods have been developed for the determination of phytosterols in vegetable oils. Such methods include HPLC with ultraviolet detection (HPLC-UV), HPLC with fluorescence detection (HPLC-FL) and liquid chromatography with tandem mass spectrometric detection (LC-MS/MS) [66-67].

Sterols of class SG and ASG can be quantitatively analyzed with gradient starting in hexane/isopropanol/water [12]. Detection is performed by a photodiode array detector. The mobile phase consisted of acetonitrile and water. The HPLC system utilized a column

Supelcosil LC-18-DB (250 x 4.6 mm) [32]. The column is maintained at 30°C in the column compartment. The flow rate of the mobile phase is 1.5 mL $min^{-1}$. Other columns for HPLC are used, ODS (4.6 x 150 mm x 5 mm) and C-30 columns. The experiments showed that steroids resolution as campesterol and stigmasterol is better using C-30 column than the ODS column [66].

In spite of the limited number of applications of UPLC in the analysis of steroids, there is growing tendency for using rapid separation techniques [67]. Recently introduced Ultra Performance Liquid Chromatography (UPLC) combined with MS analysis allows very sensitive with the minimum concentrations of phytosterols. The method UPLC-MS-APCI involves a UPLC system with binary pump and PDA detector. The column used is a C18 (reverse-phase). The solvent gradient is methanol and water with formic acid at a flow rate of 0.3 mL $min^{-1}$. The column is maintained at a constant temperature of 40°C [40].

## 5.7. Detection of Phytosteroids Mass Spectrometry

Thus, knowing the value of m/z of a molecule is possible to infer their chemical composition, and thereby determining their structure. Mass spectrometry may be used in both quantitative and qualitative analysis but it has been highlighted in the identification of mixtures of compounds, and especially in structural characterization of unknown compounds, which can be achieved by forming íons- molecule and their respective ions-fragments. We could then consider that mass spectrometry provides accurate information about the mass of the analyzed compounds. Mass spectrometry has become increasingly indispensable in the structural characterization of phytosteroids. Because of its high sensitivity and fast analysis, it is the technique of choice in analytical processes [30, 68-69].

Among the common applications of mass spectrometry tandem analysis in diagnostics is the steroid profiling in complex matrices, since in steroids, the potential for interference is greater than that for other classes of molecules, due to the presence of a large number isomers or isobars. This is even more difficult in the case of steroids that are present in low concentrations. The literature mentions the use of triple quadrupoles using different types of ionization methods may be used to generate ions such as electrospray (ESI), atmospheric pressure chemical ionization (APCI) and atmospheric pressure chemical photoionization (APPI) for analysis of steroids [68, 69]. Direct infusion modes may also be used with the addition of internal standard, for example. Multiple reaction monitoring (MRM) acquisition mode is the most used since it is a more sensitive method for the quantification of steroids. An accurate mass measurement capability with good sensitivity can also be achieved in analysis of steroids by using other TOF analyzer as Orbitrap and QTRAP [68].

Mass spectrometry has been successfully applied in recent years to a wide variety of analytical problems and composition. The mass spectrometer alone or in combination with other analytical methods has become a valuable research tool for steroids. The relative intensities of specific peaks indicate fragments particularly in the B and C rings of the steroid structure. The main fragments generated by fragmentation by mass spectrometry is dependent on factors which include the nature of the substitution at C-3, unsaturation in ring B and the type of substituent at C-17 (view Figure 1) [70].

Tandem mass spectrometric methods with triple quadrupole electrosray ionization source (ESI-MS/MS) are used for the identification and quantification of the steroids allowing high sensitivity, specificity and mass resolution [37]. The masss with source of atmospheric pressure chemical ionization (APCI) can provide structural information about the SG and ASG. Other combinations of sources and mass analyzers (e.g., electrospray coupled to MALDI-TOF) could potentially be used for analyzing intact phytosterols and their conjugates [12].

Some studies show ESI and APCI ionization modes are used. Spectra are obtained at mass range of m/z 150 to 450 Da. It is observed that the negative ion detection mode is better for the conjugated steroids (ES, SG and ASG), but is less sensitive to free steroids (FS) as compared to the positive ion mode. The sensitivity of the method using ESI is improved with low concentration or no concentration of acetic acid or ammonium acetate in the mobile phase. The same occur to the APCI ionization mode. These results suggested that the use of ammonium might acetate and acetonitrile in the mobile phase not be fine for the detection of steroids, not producing the best results when high sensitivity is required [34].

The mass spectrometer showed a tendency to show lower sensitivity to ASG containing 18: 2 (linoleic acid). The main fatty acid found in the ESG soybeans are 16: 0 (palmitic acid) comprising more than 50% of the total, followed by 16: 1 (palmitoleic acid), 18: 0 (stearic), 18: 1 (oleic), 18 : 2 (linoleic) and 18: 3 (linolenic) [32]. Analyzed FS, ES, SG and ASG via Q-TOF MS/MS operated in the positive mode by direct infusion by nanospray. Free steroids, however, have been poorly ionized after fragmentation. The results indicated that the methods ESI mass spectrometry are very sensitive to SG and ASG. The ion suppression is observed for FS and SE, while SG and ASG are less affected. Because of its low polarity, SE are susceptible to ion suppression, particularly when it has a saturated fatty acid, while SE with unsaturated fatty acids are less affected. The method developed can be easily transferred to other tandem mass spectrometers Wewer, et al., [39].

Some studies have developed a sensitive, fast and versatile method for the simultaneous analysis of twelve steroids in plants extracts by UPLC-ESI-MS/MS in positive mode including immunoaffinity chromatography (IAC) in the sample purification step. Optimization of UPLC-ESI-MS/MS for effective separation of steroids and similar hydrophobicity molecular weight is achieved by using UPLC short reverse phase column. In general, the detection limits of some steroids are reduced by additions millimolar formic or acetic acid Simersky et al., [38].

## 5.8. Direct Analysis of Phytosteroids

Although direct saponification of samples, instead of using a total lipid extract, may be used to sterol analysis, the latter offers some advantages. A total lipid extract allows a larger sample size, which can compensate for lower homogeneity of some samples and flexibility in optimizing the final concentrations of analytes GC [62].

The phytosterols can be extracted by vortex with hexane and water. The upper hexane phase is transferred to vial and maintained in an incubator at 37.5°C for 12 h, solubilized and derivatized with pyridine, hexamethyldisilazane and trimethyl-chlorosilane are added to the it is dried sample [30].

Other methods analyzed the phytosterols with hexane added in the sample and then methanol. After addition of each reagent, the mixture is vortex for 5 min. The mixture is then centrifuged to aid in phase separation. The methanolic layer containing the steroids is then removed and used for analysis. The results of the analysis indicated that both hexane and methanol are effective in the extraction of steroids [7].

### 5.9. Others Methods for Analysis of Phytosterols

The application of the near infrared spectroscopy (NIRS) could provide a good alternative for a rapid, non-destructive and low-cost estimation of phytosteroid content. However, there is no report on the development of NIRS calibrations for content phytosteroids [26, 11].

Supercritical Fluid Chromatography (SFC) is used in the analysis of steroid using carbon dioxide has a solvating power suitable for the separation of various types of steroids in the column without the need of adding modifiers. Therefore, it is possible to separate the sterols in complex samples at temperatures below theses used in GC and in shorter times than LC. The most important properties affecting SFC chromatographic separation of steroids are related to the high solvating power and low viscosity, which produces resolving power and faster throughput. In addition to its other advantages, the SFC has the ability to resolve complex mixtures of low volatility compounds allowing the direct injection of samples containing steroid with little or no pre-treatment [55].

Nuclear magnetic resonance (NMR) is reported to be a valuable tool for the identification of chemical phytosteroids since each of the carbon atoms in the molecule could potentially be examined individually proved to be essential for structural elucidation and quantification of these components [12].

## 6. Phytosterols Distribution in Different Oils

As discussed above, the determination of phytosterols in vegetable oils is of major importance for the food industry. The main methods of analysis of these steroids are mentioned, including methods of direct analysis, extraction methods and separation methods such constituents in their main forms, free and bound [32]. This section will show some qualitative and quantitative results with consideration to the analysis of phytosterols composition in oils obtained by various researchers [17, 35].

Considering the range of contents of individual phytosterols are required several calibration curves for quantification of steroids by external standardization. Thus, are you cannot always get all the diversity of steroid standards, even for many non-marketed, it uses more internal than external standardization. As internal standards, are used steroids not present in the composition vegetable oils such as cholesterol and ergosterol, or uses alcohols triterpene derivatives, such as betulin. The concentration of the different is therefore expressed in mg of sterol per kg of oil. It identification by pattern analysis or literature data under the same analysis conditions [7, 22, 41].

Table 1 shows results of the scientific literature on the types and levels of the main phytosterols in oils. The results of several different authors converge in content phytosterols, and the data shown in Table 1 is a summary of what is already on such reviews [7, 23, 35, 37]. Other results also show that the olive oil, sunflower oil and peanut oil containing mainly esterified steroids; as soybean oil contains mostly free steroids [57]. Some studies mentioning also that the corn and canola oils are very rich in esterified phytosterols (56-60%), whereas most other vegetable oils (soybean, sunflower and palm) esterified steroid contains content very lower (25-30%) [35]. For canola oil, for example, the free phytosterol fraction equal to 65% and esterified steroid fraction is 35%, the total content of phytosterols [26].

**Table 1. Total steroids levels quantified in the samples**

| Steroids | Total steroids in oils (mg $Kg^{-1}$) | | | | | |
|---|---|---|---|---|---|---|
| | β-sitosterol | Campesterol | Stigmasterol | Brassicasterol | Δ5-avenasterol | Total |
| Babassu | 380 | 100 | 60 | 10 | 110 | 660 |
| Canola | 1420 | 750 | 60 | 190 | 80 | 2500 |
| Castor | 1360 | 230 | 520 | - | 610 | 2720 |
| Coconut | 485 | 78 | 125 | - | - | 668 |
| Corn | 9680 | 2120 | 530 | - | 650 | 12980 |
| Cotton | 2590 | 180 | 20 | 80 | 170 | 3040 |
| Crambe | 1740 | 710 | 50 | 340 | 170 | 3010 |
| Moringa | 3390 | 1380 | 1680 | 20 | 440 | 6910 |
| Olive | 640 | 40 | 35 | - | 20 | 735 |
| Palm | 1510 | 530 | 250 | 40 | - | 2330 |
| Palm kernel | 750 | 90 | 100 | 30 | 20 | 990 |
| Peanut | 2110 | 440 | 170 | 10 | 320 | 3050 |
| Sesame | 2890 | 615 | 290 | 565 | 40 | 4400 |
| Soybean | 1680 | 530 | 610 | 20 | 50 | 2890 |
| Sunflower | 2190 | 280 | 270 | - | 90 | 2830 |
| Walnut | 1360 | 80 | - | - | - | 1440 |

- not detected.

The most abundant plant sterol content of different oils are observed in β-sitosterol (60%), followed campesterol and stigmasterol. Other phytosterols as campesterol, Δ5-avenasterol, Δ7-stigmasterol and Δ7-avenasterol are identified and quantified at very low levels [35]. Some publications claim that the presence of esterified phytosterols in different vegetable oils are mainly formed by oleic and linolenic fatty acids, which are part of the formation of fatty acids from most vegetable oils [35, 38]. Figure 3 illustrates the structure for the major phytosterols present in vegetable oils.

Results for analysis of phytosteroids by mass spectrometry are also important. Each steroid class has a characteristic pattern of fragmentation, and this facilitates the identification of new steroids, or even identification without the presence of patterns of reference [16, 24]. Table 2 shows only the molecular ions obtained experimentally for the main steroids presents in each class, both free and conjugated. Free steroids, for example, easily lose its side chain to form the fragment m/z 255. For the first esterified phytosterols, material loss is the fatty acid molecule, generating the fragment m/z 396. To the glycosylated phytosterols fragment m/z 396 is the first to be formed, but the loss refers to glucose molecule attached to the steroid chain. Finally, to acylated phytosterol glycosides, the major fragment to be formed when there is loss of fatty acid chain, forming fragment m/z 575 [66, 70].

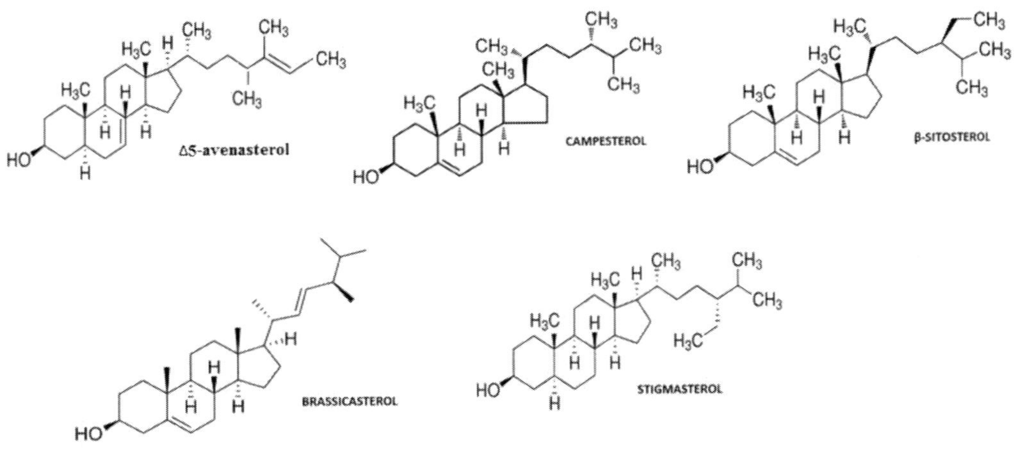


Figure 3. Illustration of the structure of the main phytosterols in vegetable oils identified.

**Table 2. Molecular ion of principal phytosterols obtained from mass spectrum**

| Steroid | Molecular Ion ($M^+$) | | | |
|---|---|---|---|---|
| | FS ($M^+$) | ES ($M^+$)[a] | SG ($M^+$) | ASG ($M^+$)[a] |
| β-sitosterol | 396 | 677 | 575 | 739 |
| Campesterol | 382 | 663 | 561 | 725 |
| Brassicasterol | 380 | 661 | 559 | 723 |
| Stigmasterol | 394 | 675 | 573 | 737 |
| Δ5-avenasterol | 394 | 675 | 573 | 737 |

[a]Molecular ion exemplifying the connection by esterified fatty acid oleic.

## Conclusion

This chapter it has as objective to the description of phytosterols, how they are distributed in oils and effective analysis methods thereof. It was possible to argue that plants synthesize a complex mixture of different types of phytosterols. The β-sitosterol is the main phytosterol, then the presence of stigmasterol, campesterol and brassicasterol. Also was discussed the importance of phytosterols with fingerprint different vegetable oils.

## References

[1] Itoh, I.; Tamura, T.; Matsumoto, T. JAOCS 1973, vol 50, 122-125.

[2] Grob, K.; Lafranchi, M.; Mariani, C. *J. Chromatogr.* 1989, vol 471, 397-408.

[3] Piironen, V.; Lindsay, D. G.; Miettinen, T. A.; Toivo, J.; Lampi, A. M. *J. Sci. Food Agric.* 2000, vol 80, 939–66.

[4] Ferrari, R. A.; Esteves, W.; Mukherjee, K. D.; Schulte, E.; *J. Agric. Food Chem.* 1997, vol 45, 4753–4757.

[5] Heinz, E. In Plant glycolipids: structure, isolation and analysis; Christie, W. W.; Ed.; Advances in lipid methodology 3. The Oily Press, Dundee, pp 211–332.
[6] Toivo, J.; Phillips, K.; Lampi, A.-M.; and Piironen, V. *J. Food Comp. Anal.* 2001, vol 14, 631-643.
[7] Chiong, D. M.; Consuegra-Rodriguez, E.; Almirall, J. R. *JFSCA* 1991, vol 37, 488-502.
[8] Bohacenko, I.; Kopicova, Z. *Gzech J. F. Sci.* 2001, vol 19, 97–103.
[9] Gordon, M. H.; Miller, L. A. D. *JAOCS* 1997, vol 74, 505-510.
[10] Weihrauch, J. L.; Gardner, J. M. *J. Am. Dietet A* 1978, vol 73, 39-46.
[11] Wojciechowski, Z. A. Biochemistry of phytosterol conjugates. In: Patterson GW, Nes WD (eds) Physiology and biochemistry of sterols. *AOCS,* pp 361–395.
[12] Moreau, R. A.; Whitaker, B. D.; Hicks, K. B. *Prog. Lipid Res.* 2002, v 41, 457–500.
[13] Kasal, A.; Budesinsky, M.; Griffiths, W. J. Spectroscopic Methods of Steroid Analysis. In: Steroid Analysis, Springer, New York, pp 27-161.
[14] Bouic, P. *J. Curr. Opin Clin. Nutr. Metab.* 2001, vol 4, 471–475.
[15] Piironen, V.; Toivo, J.; Puupponen-Pimia, R.; Lampi, A. M. *J. Food Sci. Agric* 2003, vol 83, 330–337.
[16] Bloch, K. E. *Crit. Rev. Biochem.* 1979, vol 7, 1–5.
[17] Jones, P. J.; Raeini-Sarjaz, M. *Nutr. Rev.* 2000, vol 59, 21–24.
[18] Johansson, A. *J. Am. Oil Chem. Soc.* 1979, vol 56, 886–889.
[19] Demel, R. A.; Kruiiff, B. *Biochim. Biophis. Acta* 1976, vol 457, 109–132.
[20] Jong, N.; Plat, J.; Mensink, R. P. *J. Nutr. Biochem.* 2003, vol 4, 362–369.
[21] Salo, P. W.; Ester, I.; Hopia, A. Phytosterols. In: Lipids for Functional Foods and Nutraceuticals. Gunstone, FD. (ed.), 7, pp 183-224.
[22] Ryan, E.; Galvin, K.; Connor, T. P. O.; Maguire, A. R.; Brien, N. M. O. *Plant Foods Hum Nutr* 2007, vol 62, 85–91.
[23] Piironen, V.; Toivo, J.; Lampi, A-M. *Cereal Chem.* 2001, vol 79, 148–154.
[24] Ling, W. H.; Jone, P. J. H. *Life Sci* 1995, vol 57, 195–206.
[25] Dong, Y.; Steffenson, B. J.; Chester, J. M. *J. Agric. food chem.* 2006, vol 54, 4121–4125.
[26] Seitz, L. M.; Mohr, H. E.; Burroughs, R.; Sauer, D. B. *Cereal Chem.* 1977, vol 54, 1207-1217.
[27] Hang, J.; Dussault, P. *Steroids* 2010, vol 75, 879–883.
[28] Weihrauch, J. L.; Gardner, J. M. *J. Am. Diet Assoc.* 1978, vol 73, 39–44.
[29] Griffin, J. F.; Duax, W. L.; Weeks, C. M. (eds) Atlas of Steroid Structure, vol 2. Plenum, New York.
[30] Fernandez, A. C.; Gonzalez, M. R. A.; Mendez, M. V. R.; Velasco, L. *Eur. J. Lipid Sci. Technol.* 2012, vol 114, 325–331.
[31] Ramadan, M. F.; Amer, M. M. A.; El-Saadany, S. S.; El-Masry, Awad, E. *Food Sci. Tech Int.* 2009, vol 15, 119–130.
[32] Ryan, E.; Galvin, K.; Connor, T. P. O.; Maguire, Brien, N. M. O. Plant *Foods Hum. Nutr* 2007, vol 62, 85–91.
[33] Gul, M. K.; Amar, S. *J. Cell Mol. Biol.*, vol 5, 71-79.
[34] Alonso, L.; Fontecha, J.; Lozada, L.; Juárez, M. *JAOCS* 1997, vol 74, 131-135.
[35] Verleyen, T., M.; Forcades, R.; Verhe, K.; Dewettinck, A.; Huyghebaert, De Greyt, W. *JAOCS* 2002, vol 79, 117–122.
[36] Johannsson, A.; Appelqvist, L. A. *Lipids* 1978, vol 13, 658-665.

[37] Jones, P. J.; Raeini-Sarjaz, M.; Ntanios, F. Y.; Vanstone, C. A.; Feng, J. Y.; Parsons, W. E. *J. Lipid Res.* 2000, vol 41, 697–705.
[38] Johannsson, A. *Lipids* 1979, vol 14, 285-291.
[39] Kolesnikova, M. D.; Xiong, Q.; Lodeiro, S.; Hua, L.; Matsuda, S. P. T. *Arch. Bioch. Bioph.* 2006, vol 447, 87–95.
[40] Grob, K.; Lanfranchi, M. *J. Chromatogr. A* 1989, vol 471, 397–405.
[41] Kovganko, N. V.; Kashkan, Z. N. *Chem. Nat. Compd.* 1999, vol 35, 479–497.
[42] Toivo, J.; Phillips, K.; Lampi, A-M.; Piironen, V. *J. Food Comp. Anal.* 2001, vol 14, 631–643.
[43] Robert, F. J.; Javitt, N. B. *Steroids* 2011, vol 76, 211–215.
[44] Beveridge, T. H.; LI, T. S.; DROVER, J. C. *J. Agric. Food Chem.* 2002, vol 50, 744–750.
[45] New Data for Cholesterol Contents in Meat, Fish, Milk, Eggs and Their Products Consumed in Finland.
[46] Piironen, V.; Toivo, J.; Lampi, A-M. *J. Food Comp. Anal.* 2000, vol 13, 619-624.
[47] Rao, A. V.; Janezic, S. A. *Nutr. Cancer* 1992, vol 18, 43-52.
[48] Jonker, D.; Hoek, G. D.; Glatz, C. H.; Homan, C.; Posthumus, M. A.; Katan, M. B. *Nutr. Rep. Int.* 1985, vol 32, 943-951.
[49] Ostlund, R. E.; Racette, S. B.; Okeke, A. *Am. J. Clin. Nutr.* 2002, vol 75, 1000–1004.
[50] Maria, T.; Robert, M. *Food Chem.* 1987, vol 25, 251–258.
[51] Frega, N.; Bocci, F.; Lercker, G. *JAOCS* 1992, vol 69, 447-450.
[52] Al-Ismail KM, Alsaed AK, Ahmad R, Al-Dabbas. D etection of olive oil adulteration with some plant oils by GLC analysis of sterols using polar column. *Food Chemistry* 2010; 121:1255–9.
[53] Jawad, I.M.; Kochhar, S.P.; Hudson, B.J.F. *Lebensm Wiss Technol.* 1984, vol 17, 155–159.
[54] Vlahakis, C.; Hazebroek, J. *JAOCS* 2000, vol 77, 49-53.
[55] Gere, D. R. *Science* 1983, vol 222, 253-261.
[56] Choong, Y. M.; Lin, H. J.; Chen, C. W.; Wang, M. L. *J. Food Drug Anal.* 1999, vol 7, 279-287.
[57] Artho, G.; Grob, K.; Marianai, C. *Science and Technology* 1993, vol 95, 176–180.
[58] Grob, K.; Lanfranchi, M.; Mariani, C. *J. Chromatogr.* 1989, vol 471, 397-405.
[59] Bhawani, S. A.; Sulaiman, O.; Hashim, R.; Ibrahim, M. N. M. *Trop. J. Pharm. Res.* 2010, vol 9, 301-313.
[60] Zarzycki, P. K.; Baran, M.; Wlodarczyk, E.; Bartoszuk, M. A. *J. Liq. Chromatogr. Relat. Technol.* 2007, vol 30, 2629- 2634.
[61] Lepage, M. *J. Lipid Res.* 1964, vol 5, 587-592.
[62] Phillips, K. M.; Ruggio, D. M.; Khorassani, M. A. *J. Food Lips* 2005, vol 12, 124-140.
[63] Soupas, L.; Juntunen, L.; Säynäjoki, S.; Lampi, A-M.; Piironen, V. *JAOCS* 2004, vol 81, 135-141.
[64] Bezerra, K. S.; Antoniosi Filho, N. R. *J. Braz. Chem. Soe* 2014, vol 2, 238-251.
[65] Bezerra, K. S.; Antoniosi Filho, N. R. *Fuel* 2014, vol 130, 149-162.
[66] Ferguson, P. L.; Iden, C. R.; McElroy, A. E.; Brownawell, B. *J. Anal. Chem.* 2001, vol 73, 3890–3895.
[67] Shackleton, C. *J. Steroid Biochem. Mol. Biol.* 2010, vol 121, 481–490.
[68] Brooks, C. J. W.; Horning, E. C.; Young, J. S. *Lipids* 1968, vol 3, 391–402.

[69] Soldin, S. J.; Soldin, O. P. *Clin. Chem.* 2009, vol 55, 1061–1066.
[70] Friedland, S. S.; Lane, G. H.; Longman, R. T.; Train, K. E. *Anal. Chem.* 1959, vol 31, 169-174.

In: Phytosterols
Editor: Deanna Garner

ISBN: 978-1-63483-477-3


*Chapter 10*

# ESTIMATION OF PLANT STEROL INTAKE IN A RESIDENT POPULATION OF THE VALENCIAN COMMUNITY (SPAIN)

***Guadalupe Garcia-Llatas, Evelyn Mejía, Antonio Cilla*, Amparo Alegría and María Jesús Lagarda***
Nutrition and Food Science Area, Faculty of Pharmacy, University of Valencia, Burjassot, Valencia, Spain

## ABSTRACT

Plant sterols (PS), including phytosterols and phytostanols, are bioactive compounds which when consumed have several beneficial effects on the human body, including antiinflammatory and anticancer activities. In addition, a daily intake of 1.5-3 g PS results in an 8-10% reduction in LDL-cholesterol levels. Previous studies in northern European countries have reported PS intakes from natural and PS-enriched sources, though no recent studies have been made in the Mediterranean area, where the diet is known to be naturally rich in PS-containing foodstuffs (e.g., olive oil, nuts, etc.). Therefore, the aim of the present study was to describe the common PS sources and daily intakes of PS in a resident population (n = 75, 50.7% men, age = 20-68 years) in the Valencian Community (a region located on the Spanish Mediterranean coast) between November 2013 and May 2014, and to compare it with data collected in the year 2005 from a similar type of survey conducted among residents of the same region (n = 70, 47.1% men, age = 20-77 years). Intake was estimated by using a food frequency questionnaire considering naturally PS-containing food products from the normal diet. The results showed total PS intake to be similar in both study periods (274 mg/day in 2013-2014 versus 282 mg/day in 2005). Although the predominant PS content in natural sources corresponded to oils followed by nuts and legumes, on considering their frequency of consumption, the main PS sources were seen to be legumes > vegetables > oils. In sum, the recorded PS intake falls within the range reported by other studies in European populations (211-325 mg/day).

---

* Corresponding author: Antonio Cilla (antonio.cilla@uv.es).

# 1. INTRODUCTION

## 1.1. Chemical Structure

Plant sterols (PS) are compounds which have three terpenes (triterpenes) and are composed of 30 carbon atoms. They differ from cholesterol in the presence of a methyl or ethyl group in the side chain (with the possible presence of an unsaturated bond), in position C-24 (Figure 1). There are two main groups of PS: phytosterols and phytostanols. The former have a double bond in position C-5, while the latter lack this bond. Phytostanols are the product of phytosterol hydrogenation, and are present in lesser abundance relative to the amount of phytosterols.

Although more than 200 types of phytosterols have been identified, the most important and also the most abundant forms are β-sitosterol, campesterol and stigmasterol. Likewise, in the case of phytostanols, β-sitostanol and campestanol are the most important molecules (García-Llatas and Rodríguez-Estrada, 2011).

Plant sterols can be found in nature in either free or conjugated form. The conjugation process can take place by esterification with fatty acids, ferulate or ferulic acid (an antioxidant compound similar to vitamin E and C) via the 3β-OH group or glycosylation. Conjugation with fatty acids is important, since it represents almost 50% of the total sterols in foods; the same applies in the case of ferulic acid, though to a lesser extent. On the other hand, PS that have undergone glycosylation constitute a minority presence in plant foods, though there are some exceptions such as potatoes. In addition, an important difference between a free sterol and a conjugated sterol is that the latter dissolves easily and quickly in high-fat foods, while free sterols do so more slowly (Palou et al., 2005).

## 1.2. Biological Effects

The possible roles of PS in the human body are varied, though the most widely studied and demonstrated functions are their capacity to reduce plasma cholesterol levels in the body, which is beneficial for reducing the risk of atherosclerosis - one of the leading causes of death due to cardiovascular disease in western countries (Awad et al., 2001). This means that PS function by inhibiting cholesterol intestinal absorption in the human body. Moreover, as hydrophobic sterols, they tend to displace cholesterol from the micelles during absorption in the intestinal epithelium, thereby favoring decreased uptake. However, it should be noted that such inhibition of absorption results in slight cholesterol deficiency - causing the production of endogenous cholesterol to increase as a partially compensating mechanism (Brüll et al., 2009).

While cholesterol is readily absorbed in the intestine (56.2% absorption), the absorption of phytosterols ranges from 0.4-3.5%, and in the case of phytostanols the amount absorbed is much smaller (0.02-0.3%). The explanation for this is not fully clear, though several studies have shown that poor absorption can be due to the low affinity of acyl-coenzyme A cholesterol transferase (ACAT) for PS. This enzyme is responsible for PS esterification. The next step comprises incorporation to chylomicrons, followed by lipidic metabolism in the liver in order to facilitate elimination. Other studies suggest that their absorption is related to the inherent structure of PS, namely the side chain and also the double bond which some of these molecules present. These are some of the factors that may influence absorption (Ostlund, 2002; Palou et al., 2005). Furthermore, poor absorption may be influenced by ABC transporters (ATP binding cassette), which are responsible for removing cholesterol and β-sitosterol to prevent absorption and accumulation (Palou et al., 2005; Rozner and Garti, 2006).

The first observations of the cholesterol-lowering effects of PS date back more than 50 years ago when a decrease in plasma cholesterol was observed on feeding chickens with soybeans. Similar effects were documented in subsequent animal studies - the effects of soy being attributed to its PS content. Interest in these molecules further increased during the 1990s, when PS esterification with fatty acids led to their incorporation in foods such as margarines and cereals, among others (Palou et al., 2005).

Besides their ability to reduce cholesterol levels, PS have other beneficial biological effects referred to cancer, membrane properties, altered endothelial function and the immune system. Several studies have indicated that PS may have health-promoting effects, such as anticancer activity (Woyengo et al., 2009). Awad et al. (Awad and Fink, 2000; Awad et al., 2001) provided evidence that phytosterols were toxic to breast cancer cells. Other studies indicated that phytosterols were toxic to colon cancers (Awad et al., 1998; Rao and Janezic, 1992). A review on the modulation of signal transduction in cancer cells by phytosterols has recently been published by the same research group (Bradford and Awad, 2010). In this work, phytosterols are defined as cancer chemopreventive dietary phytochemicals, since they increase sphingomyelin turnover, ceramide formation and liver X receptor activation. Consequently, these actions slow cell cycle progression, inhibit cell proliferation and activate caspase cascades and apoptosis in cancer cells. Paniagua-Pérez et al. (2008) determined the antigenotoxic capacity and lymphocyte induction potential in mice, as well as the *in vitro* capacity of β-sitosterol to trap free radicals - thus showing its potential as a chemopreventive agent. Recently, other effects referred to endothelial function, oxidative stress and antioxidant status, coagulation and platelet aggregation, inflammation, neurocognitive function, eye disease and anti-osteoarthritic actions have been reviewed (Brüll et al., 2009; Jones and AbuMweiss, 2009; Derdemezis et al., 2010; Rudkowska, 2010; Gabay et al., 2010).

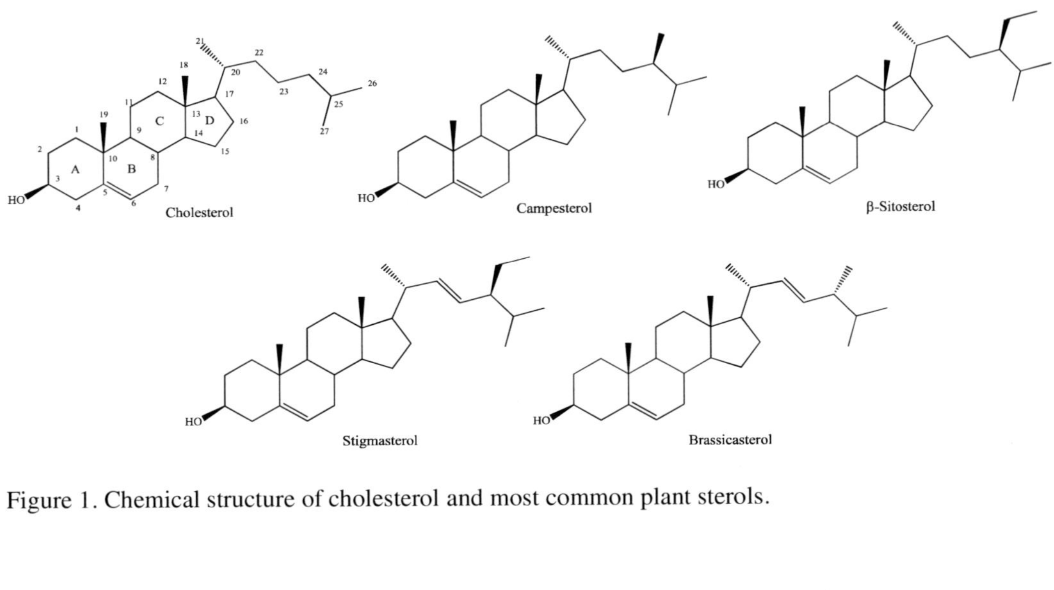


Figure 1. Chemical structure of cholesterol and most common plant sterols.

However, these compounds can have side effects, since they may decrease the absorption of carotenoids and other fat soluble vitamins (vitamins E, K and D). This situation needs to be considered in cases where there is a deficiency of these vitamins, or in physiological situations where the requirements are increased, such as during pregnancy, lactation or infancy (Rozner and Garti, 2006, Christos et al., 2010). It is therefore appropriate to recommend a higher intake of fruits and vegetables in a diet containing PS-enriched products, to compensate for the decreased bioavailability of both β-carotene and other fat-soluble nutrients in such situations.

## 1.3. Natural Food Sources

As previously mentioned, more than 200 different PS have been identified, β-sitosterol, followed by campesterol and stigmasterol being the most abundant in foods. In the case of phytosterols, these are mainly found in fruits, vegetables, seeds, grains and nuts as natural sources of PS. On the other hand, despite the few studies that have been made of phytostanols, plants represent their main natural source, though in smaller amounts compared to phytosterols. Furthermore, these compounds can be chemically obtained by hydrogenating phytosterols.

The highest concentrations of PS in most representative rich foods have been reported in vegetable oils such as rice bran oil, wheat germ oil and corn oil, with values ranging from 900-1050 mg/100 g. These foodstuffs are followed by nuts (104-360 mg/100 g), cereals (20-344 mg/100 g) and legumes (124-135 mg/100 g). The lowest quantities are found in fruits and vegetables (1-50 mg/100 g) (Palou et al., 2005; Marangoni et al., 2010; Bacchetti et al., 2011). Therefore, considering the recommendations for reaching a hypocholesterolemic effect (1.5-3 g/day), it is clear that even a balanced diet is not enough to achieve the optimum estimated levels for maximum PS effect.

## 1.4. Legislation

In the year 2004, the European Commission (EC, 2004) indicated that, based on the available information, the cholesterol-lowering effect of PS is manifest from values of 1.5 g to 3 g PS/day, as has also been shown in several studies (Jakulj et al. 2013; Ras et al. 2013). But this amount of PS depends on the food source involved, the PS formulation used in fortified foods, and the dietary habits of the population, among other factors. This European Commission report also noted that a maximum daily consumption of 3 g of PS allows a 12% decrease in LDL-cholesterol, as confirmed by other studies such as that published by Hansel et al. (2011), showing serum LDL-cholesterol and total cholesterol reductions of 13% and 10%, respectively - thereby facilitating a decrease in cardiovascular events. In line with this, in 2012 the European Food Safety Authority (EFSA) declared that a daily intake of 3 g of plant sterols or stanols lowers LDL-cholesterol with similar efficacy (11.3%), and the effect is achieved when consumption is continued for at least 2-3 weeks (EFSA, 2012). However, it has not been ruled out that values above 3 g of PS a day might display more hypocholesterolemic effects, or also cause undesirable effects related with to decreased absorption of β-carotene (provitamin A) and other fat-soluble nutrients (EC, 2004).

Health claims have recently been approved for these functional foods in the European Union (Regulations 983/2009/EC, 432/2012/EC and 686/2014/EC), referred to the beneficial effects of phytosterols and phytostanols in managing blood cholesterol levels. Moreover, the European Atherosclerosis Society (EAS) Consensus Panel, based on data referred to the lowering of LDL-cholesterol and the absence of adverse signs (associated with a PS intake of 2 g/day), concluded that functional foods with PS may be considered in individuals with high cholesterol levels at intermediate or low global cardiovascular risk who do not qualify for drug treatment, as an adjunct to therapy in high and very high risk patients who fail to achieve LDL-cholesterol targets with statins or who are statin-intolerant, and in adults and children (over 6 years of age) with familial hypercholesterolemia (Gylling et al., 2014).

Different studies on PS intake from natural sources point to differences between Western and Eastern diets. The former is associated to reported values of 200-300 mg PS/day, whereas the intake values are higher in Eastern diets (200-500 mg PS/day), with levels close to those obtained in vegetarian diets. Furthermore, the higher the diet PS content, the lower the cholesterol absorption and plasma cholesterol values (Espín and Tomás-Barberán, 2005).

## 1.5.European Studies on Estimated Plant Sterol Intake

Several studies have been carried out in Europe (Belgium, Sweden, Germany, United Kingdom, Spain, Ireland, Finland and the Netherlands) on PS intake from both natural (Normén et al., 2001; Valsta et al., 2004; Jiménez-Escrig et al., 2006; Klingberg et al., 2008; Hearty et al., 2009; Sioen et al., 2011; Klingberg et al., 2012; Marttinen et al., 2014) and enriched sources (Kuhlmann et al., 2005; Hearty et al., 2008; Marttinen et al., 2014) (Table 1).

Regarding the studies on PS intake from natural food sources, four of them have used 24 or 48 hour dietary recall as data collection method, while 5 of them have made use of a food frequency questionnaire (FFQ) to estimate PS intake. The minimum mean intake of PS was 211 mg/day in Sweden (Klingberg et al., 2012), and the maximum mean intake was 325 mg/day in Finland (Marttinen et al., 2014). In terms of gender, the lowest values for men (226 mg/day) and women (197 mg/day) were also reported in Sweden (Klingberg et al., 2012), whereas the maximum levels in men (363 mg mg/day) were reported in Finland (Marttinen et al., 2014) and the maximum levels in women (293 mg/day) corresponded to the United Kingdom (Klingberg et al., 2008). The main dietary sources of PS in the populations under study (obtained from both local and foreign food composition databases) were bread, cereals and vegetable oils.

On the other hand, studies on PS intake considering enriched food sources have been conducted using three different methods: a simulated prospective method with model PS fortified foods (Kuhlmann et al., 2005); an interview-assisted questionnaire (Hearty et al., 2008); and a 48 hour dietary recall protocol (Marttinen et al., 2014). The highest and lowest mean intakes of PS were observed in Germany using the simulated prospective method, with values of 7000 mg/day and 1500 mg/day, respectively. Sexing terms of gender, men in Ireland showed the maximum PS intake values (2710 mg/day) (Hearty et al., 2008), while women in Finland showed the minimum PS intake values (1600 mg/day) (Marttinen et al., 2014) among the different studies involving PS-enriched foods. The main sources of these fortified foods were dairy products, liquid yoghurt and margarine.

**Table 1. Different studies carried out in Europe about the intake of plant sterols. Adapted from Normén et al., 2001, Valsta et al., 2004, Jiménez-Escrig et al., 2006, Klingberg et al., 2008, Hearty et al., 2009, Sioen et al., 2011, Klingberg et al., 2012, Marttinen et al., 2014 for natural food sources; Kuhlmann et al., 2005, Hearty et al., 2008, Marttinen et al., 2014 for enriched food sources.*Food frequency questionnaire (FFQ)**

| Countries | | Population | | | Method Used | Intake (mg PS/day) | | | Main Source | | Others |
|---|---|---|---|---|---|---|---|---|---|---|---|
| | | Total sample | Women | Men | | Mean | Men | Women | Natural foods | Foods enriched with PS | |
| Natural sources | Netherlands (Normén et al., 2001) | 3123 | 1525 | 1598 | Semi-quantitative FFQ | 285 | 307 | 262 | Bread, vegetable oils | - | Relationship between intake and cancer |
| | Finland (Valsta et al., 2004) | 2738 | 1446 | 1292 | 24h dietary recall | 271 | 305 | 237 | Bread, cereals | - | - |
| | Spain (Jiménez-Escrig et al., 2006) | 6000 | - | - | FFQ | 276 | - | - | Cereals, vegetable oils | - | Assessment of PS content in foods |
| | United Kingdom (Klingberg et al., 2008) | 24789 | 13571 | 11227 | FFQ | 296.5 | 300 | 293 | Cereals | - | - |
| | Ireland (Hearty et al., 2009) | 1379 | 662 | 717 | FFQ | 254 | 282.7 | 227.5 | Bread | - | - |

**Table 1. (Continued)**

| Countries | | Population | | | Method Used | Intake (mg PS/day) | | | Main Source | | Others |
|---|---|---|---|---|---|---|---|---|---|---|---|
| | | Total sample | Women | Men | | Mean | Men | Women | Natural foods | Foods enriched with PS | |
| Natural sources | Belgium (Sioen et al., 2011) | 3038 | 1523 | 1538 | 24h dietary recall | 265 | 301 | 229 | Bread, cereals | - | - |
| | Sweden (Klingberg et al., 2012) | 203 | 101 | 102 | FFQ | 238 | 257 | 219 | Cereals | - | - |
| | | | | | 24h dietary recall | 211 | 226 | 197 | Cereals | | |
| | Finland (Marttinen et al., 2014) | 194 | 88 | 106 | 48h dietary recall | 325 | 363 | 286 | - | - | - |
| Enriched sources | Germany (Kuhlmann et al., 2005) | 23209 | - | - | Simulated prospective method with three kind of PS enrichment (2g, 1-2g and 0,3g) | Up to: 1st:7000 2nd 2500 3rd:1500 | - | - | - | Dairy products | Prospective simulation of PS intake |
| | Ireland (Hearty et al., 2008) | 468 | 186 | 282 | Interview-assisted questionnaire | 2450 | 2710 | 2290 | - | Liquid yoghurt | - |
| | Finland (Marttinen et al., 2014) | 194 | 88 | 106 | 48h dietary recall | 1900 | 2200 | 1600 | - | Margarine | - |

## 2. Aim of the Study

The main objective of the present study was to estimate PS intake in a resident population in the Valencian Community (a region located on the Spanish Mediterranean coast) through the administration of a food frequency questionnaire. The data obtained in turn were used to establish comparisons with the results of a similar survey performed in the same region in 2005.

## 3. Methodology

Food intake was evaluated by means a food frequency questionnaire in the period from November 2013 to May 2014. The survey contained data on gender, age, education, lifestyle, medication and a list of food items containing PS, distributed into 8 food groups (nuts, oils and fats, cereals, chocolates, legumes, vegetables, fruits and others), and frequency of intake. In addition, individuals in the study were provided with the printed survey, and were given a brief explanation of its structure (food groups, portions, etc.), with instructions on how to complete the questionnaire. Plant sterol total contents as well as β-sitosterol, campesterol and stigmasterol in foods included in the food frequency questionnaire were compiled from data previously reported by different authors (Table 2).

**Table 2. Plant sterol contents (mg/g) in the food items included in the food frequency questionnaire. (Normén et al., 1999; Souci, 2000; Normén et al., 2002; Jiménez-Escrig et al., 2006; Normén et al., 2007; Lama-Muñoz et al., 2011)**

| Food groups | mg PS / g product | β-sitosterol (mg/g) | Campesterol (mg/g) | Stigmasterol (mg/g) |
|---|---|---|---|---|
| **Nuts** | | | | |
| Almond | 1.480 | 1.255 | 0.044 | - |
| Cashew | 1.580 | 1.050 | 0.030 | - |
| Hazelnut | 1.280 | 1.109 | 0.069 | - |
| Peanut | 1.430 | 0.994 | 0.190 | - |
| Roasted corn | 0.700 | - | - | - |
| Nut | 1.700 | 1.024 | - | - |
| Pine kernel | 1.310 | 1.002 | 0.053 | - |
| Sunflower seed | 2.260 | 1.464 | 0.227 | - |
| Pistachio | 2.420 | 2.019 | 0.122 | - |
| **Oils and fats** | | | | |
| Refined olive oil | 2.359 | 1.210 | 0.062 | - |
| Virgin olive oil | 2.597 | 1.585 | 0.063 | - |
| Pomace oil | 2.785 | 2.495 | 0.900 | 0.23 |
| Corn oil | 9.090 | 5.860 | 2.150 | 0.66 |
| Sunflower oil | 4.925 | 2.193 | 0.364 | - |
| Margarine | 2.170 | 1.220 | 0.470 | - |
| **Cereals** | | | | |
| Corn | 0.220 | 0.170 | 0.043 | - |
| Mix | 0.730 | 0.420 | 0.150 | 0.024 |

| Food groups | mg PS / g product | β-sitosterol (mg/g) | Campesterol (mg/g) | Stigmasterol (mg/g) |
|---|---|---|---|---|
| White wheat bread | 0.419 | 0.261 | 0.077 | - |
| Wholemeal bread | 0.700 | 0.440 | 0.140 | 0.017 |
| Maryland cookies | 0.670 | 0.440 | 0.190 | 0.048 |
| Sweet crackers | 0.490 | 0.300 | 0.140 | 0.016 |
| Chocolate biscuits | 0.700 | 0.435 | 0.083 | - |
| Bagel | 0.880 | 0.450 | 0.150 | - |
| Rice | 0.290 | 0.198 | 0.046 | - |
| Pasta | 0.280 | 0.190 | 0.040 | - |
| **Chocolate** | | | | |
| Chocolate buns | 0.330 | 0.134 | 0.130 | 0.043 |
| Chocolate | 1.260 | 0.776 | 0.123 | 0.314 |
| **Legumes** | | | | |
| Lentil | 1.173 | 0.808 | 0.100 | 0.108 |
| Chickpea | 1.212 | 0.841 | 0.124 | 0.083 |
| White bean | 1.081 | 0.563 | 0.040 | 0.352 |
| **Vegetables** | | | | |
| Chard | 0.166 | 0.069 | 0.009 | 0.0053 |
| Spinach | 0.163 | 0.163 | 0.005 | 0.081 |
| Artichoke | 0.485 | 0.169 | 0.019 | 0.074 |
| Celery | 0.078 | 0.029 | 0.007 | 0.041 |
| Leek | 0.117 | 0.045 | 0.009 | 0.059 |
| Eggplant | 0.059 | 0.032 | 0.004 | 0.016 |
| Marrow | 0.024 | 0.012 | 0.001 | 0.006 |
| Pumpkin | 0.120 | 0.120 | - | - |
| Onion | 0.072 | 0.026 | 0.013 | 0.003 |
| Champignon Mushroom | 0.180 | - | 0.180 | - |
| Cabbage | 0.274 | 0.176 | 0.047 | 0.016 |
| Cauliflower | 0.183 | 0.278 | 0.111 | 0.043 |
| Green asparagus | 0.106 | 0.054 | 0.022 | 0.010 |
| Pea | 1.35 | 0.600 | 0.750 | - |
| Lima bean | 1.240 | 0.520 | 0.720 | - |
| Green bean | 0.188 | 0.093 | 0.012 | 0.047 |
| Lettuce | 0.135 | 0.073 | 0.034 | 0.009 |
| Endive | 0.169 | 0.077 | 0.016 | 0.061 |
| Escarole | 0.181 | 0.131 | 0.010 | 0.028 |
| Roast potato | 0.043 | 0.034 | 0.001 | 0.006 |
| French fries | 0.240 | 0.100 | 0.040 | 0.100 |
| Cucumber | 0.070 | 0.003 | 0.001 | 0.015 |
| Red pepper | 0.074 | 0.039 | 0.013 | 0.002 |
| Green pepper | 0.094 | 0.056 | 0.019 | 0.004 |
| Mushroom | 0.180 | 0.180 | - | - |
| Tomato | 0.099 | 0.035 | 0.008 | 0.021 |
| Carrot | 0.186 | 0.126 | 0.027 | 0.031 |
| Garlic | 0.182 | 0.026 | 0.0027 | - |
| **Fruits** | | | | |
| Orange | 0.304 | 0.220 | 0.038 | 0.012 |
| Grapefruit | 0.170 | 0.130 | 0.020 | 0.013 |

**Table 2. (Continued)**

| Food groups | mg PS / g product | β-sitosterol (mg/g) | Campesterol (mg/g) | Stigmasterol (mg/g) |
|---|---|---|---|---|
| Banana | 0.201 | 0.133 | 0.018 | 0.130 |
| Apple | 0.160 | 0.138 | 0.008 | 0.023 |
| Pear | 0.110 | 0.106 | 0.004 | 0.003 |
| Strawberry | 0.113 | 0.082 | 0.007 | 0.002 |
| Cherry | 0.201 | 0.162 | 0.011 | 0.005 |
| Plum | 0.189 | 0.150 | 0.016 | 0.007 |
| Peach | 0.146 | 0.088 | 0.004 | 0.025 |
| Apricot | 0.152 | 0.059 | 0.008 | 0.019 |
| Watermelon | 0.045 | 0.024 | 0.006 | 0.004 |
| Melon | 0.046 | 0.023 | 0.003 | 0.008 |
| Kiwi | 0.071 | 0.049 | 0.002 | 0.007 |
| Grape | 0.201 | 0.228 | 0.021 | 0.003 |
| **Others** | | | | |
| Mayonnaise | 4.840 | 2.510 | 1.810 | - |
| Olives | 0.425 | 0.410 | 0.018 | - |

The surveyed population in 2013-2014 included 75 people (38 men and 37 women) aged between 20-68 years (mean 36.2 ± 14.9). In comparison, the PS intake estimation in 2005 involved 70 people (33 men and 37 women) aged between 20-77 years (mean 39.7 ± 11.9). All subjects were informed and gave their written consent.

The Dyane software package (version 4) (Madrid, Spain) was used to collect data from the questionnaires in different categories and to evaluate PS intake.

# 4. Results and Discussion

## 4.1. Food Frequency of Intake

Results related to food frequency of intake, expressed as median g/person/year, and the contribution made by each food group are shown in Table 3.

On comparing the results obtained in both surveys, a higher total food intake was recorded in 2005 compared with the period 2013-2014 (427 vs. 256 kg/person/year). The main percentage contribution to food intake corresponded to vegetables, with values of 64.11% and 49.02% in 2005 and 2013-2014, respectively. In the most recent survey, the highest intake within this food group corresponded to raw tomato, lettuce and onion, while the lowest intakes corresponded to pumpkin, green asparagus, endives and celery. In the case of the 2005 survey, the most consumed vegetables were raw tomato, artichoke and onion, while celery, escarole and garlic were the least consumed.

After vegetables, fruits were the second most consumed foods, with values of 23.53% (2013-2014) and 21.58% (2005). In the latest survey, the highest values corresponded to orange and apple, while grapefruit, grapes and apricots were the least consumed fruit items. In the 2005 survey, oranges and apples were likewise the most consumed fruits, while grapefruit, plum and apricot showed the lowest percentages.

In both surveys, the third highest intake food group was represented by legumes, with percentage contributions of 11.7% (2013-2014) and 6.18% (2005). Lentils were the most consumed item and white beans the least consumed. Cereals appeared next on the list of most consumed foods, with percentage contributions of 10.91% (2013-2014) and 4.75% (2005). White wheat bread, rice and pasta were the most consumed in 2013-2014, and corn cereals, mix and sweet crackers the least consumed. On the other hand, white wheat bread, rice and pasta were the most consumed foods in 2005, while the least consumed corresponded to corn and mix cereals, Maryland cookies, sweet crackers and chocolate biscuits.

Oils and fats were the fifth group in order of consumption (2.48% in 2013-2014 and 1.71% in 2005). In both surveys, refined olive oil and virgin olive oil were the most consumed items, and olive pomace oil and corn oil the least consumed. The category "Others" followed next on the list, with percentage contributions of 1.21% and 0.34% in 2013-2014 and 2005, respectively. Olives and mayonnaise were respectively the most and least consumed food items in this category.

In seventh place, nuts contributed 0.63% to the total consumption in both surveys, with almonds, peanuts and nuts as the most consumed, and pistachios and sunflower seeds (2013-2014), and cashew nuts and roasted corn (2005) as the least consumed. Finally, the lowest ranking food group corresponded to chocolate, with percentage contributions of 0.61% (2013-2014) and 0.7% (2005).

## 4.2. Plant Sterols Intake

Plant sterols intake in food items selected for the food frequency questionnaire expressed as mg/day and percentage contribution are shown in **Table 4**. The intake of PS in the population under study was influenced not only by the content of PS in the foods consumed but also by the frequency of intake of the selected food items in the food frequency questionnaire.

The highest PS intakes corresponded to the group of legumes in both 2013-2014 (35.26%) and 2005 (30.04%), lentils being the food item supplying the highest intake. The second most important group corresponded to vegetables, with 19.27% (2013-2014) and 20.05% (2005). In the last survey, lima bean, raw tomato and carrots supplied the highest PS values, while in the first survey the highest contributions corresponded to artichoke, raw tomato and lima bean. The third position corresponded to oils and fats in 2013-2014 (16.14%) - virgin olive oil and refined olive oil being the food items with the greatest contributions. However, in the survey of 2005, the third ranking group corresponded to fruits (18.23%), with oranges and apples as the main contributors. The fourth group in the survey of 2013-2014 corresponded to fruits (11.28%) with orange, banana and apples as the main suppliers, while in 2005 this position corresponded to oils and fats (6.85%) - virgin olive oil yielding the highest values. In both 2013-2014 (10.33%) and in 2005 (6.85%) cereals represented the fifth most important group referred to intake, with white wheat bread as the leading supplier, followed by rice and pasta. The category "Others" ranked in sixth place (3.34%) in the period 2013-2014, with the predominant contribution of mayonnaise (a product mainly composed of oil). However, in the survey of 2005, nuts (3.87%) ranked in this position due to the contribution to PS intake of almonds and peanuts. In turn, nuts (2.42%) were reported in the seventh position in 2013-2014, with almonds and peanuts as the main representatives. In the case of the 2005 survey, chocolate (1.76%) held this position. Finally, chocolate occupied the last position (1.96%) in the 2013-2014 survey, while the category "Others" (1.62%) ranked in this place during 2005 - with mayonnaise as main supplier of PS.

As it can be seen in Table 4 and Figure 2, the total PS intake was very similar in both surveys (274 mg/day in 2013-2014 vs. 282 mg/day in 2005). On comparing the results of the present study with those previously reported for Spain by Jiménez-Escrig et al. (2006), the values of the latter study were seen to be similar to those recorded in the Valencian Community, which is therefore representative of Spain as a whole. It must be underscored, however, that all these results may be influenced by different parameters such as season, geographic region and gender, among others.

On considering the distribution among food groups (Figure 2), four dominant groups were observed in 2013-2014, with total contributions to PS intake greater than in the survey of 2005 (legumes, cereals, "Others" and chocolate, with 96, 28, 9 and 5 mg/day). In contrast, the first survey of 2005 underscored higher PS intakes in the food groups corresponding to vegetables, fruits, oils and fats and nuts (56, 51, 50 and 11 mg/day) versus the more recent survey of 2013-2014.

## 4.3. Comparison with European Studies on Plant Sterols Intake

In order to compare the values of the present study carried out in the Valencian Community with others recorded in Europe, Table 1 summarizes different studies conducted in different European countries in recent years. As can be seen, each country reported different values referred to PS intake. This fact can be due not only to the year when the study was completed, the seasons involved or the objectives of the study, but also to the samples used and the geographic region involved.

On comparing food sources in the different studies, great differences between natural and enriched food sources were observed, as expected. In the first case, the mean intake was 269 mg/day for natural sources, while far higher values were recorded for enriched sources, with a mean intake of 3070 mg/day.

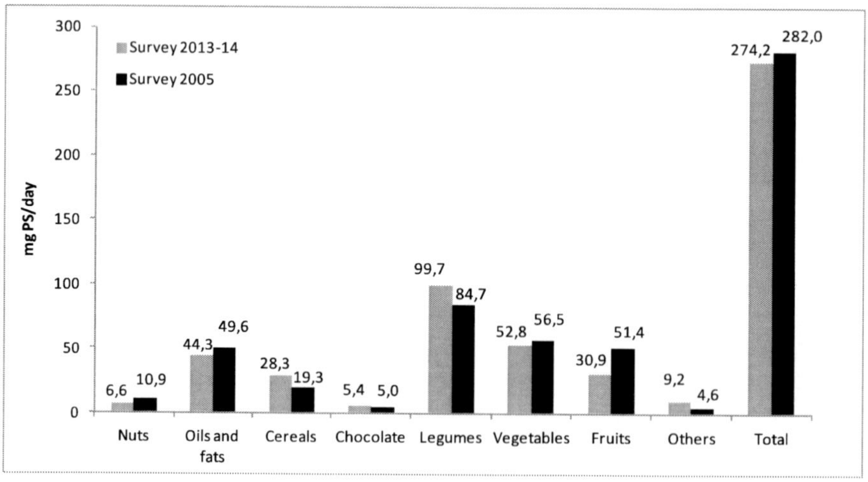


Figure 2. Comparison of PS intake values (mg/day) obtained in 2013-2014 and 2005 surveys depending on the different food groups.

**Table 3. Food frequency intakes obtained in 2013-2014 and 2005 surveys, expressed as median in g/person/year, and the contribution in percentage from each food group**

| | Intake frequency | | | |
|---|---|---|---|---|
| | Survey 2013-2014 | | Survey 2005 | |
| Food group | Median: g/person/year (min-max) | Contribution (%) | Median: g/person/year (min-max) | Contribution (%) |
| **Nuts** | | | | |
| Almond | 480 (0-14600) | 0.63 | 960 (0-14600) | 0.63 |
| Cashew | 0 (0-2080) | | 0 (0-2080) | |
| Hazelnut | 20 (0-14600) | | 120 (0-1870,26) | |
| Peanut | 480 (0-62400) | | 960 (0-14600) | |
| Roasted corn | 0 (0-10400) | | 0 (0-4160) | |
| Nut | 480 (0-58400) | | 480 (0-29200) | |
| Sunflower seed | 75 (0-10950) | | 30 (0-10950) | |
| Pistachio | 80 (0-14600) | | 120 (0-14600) | |
| Sum of Group | 1615 | | 2670 | |
| **Oils and fats** | | | | |
| Refined olive oil | 2600 (0-14600) | 2.48 | 3650 (0-25550) | 1.71 |
| Virgin olive oil | 3650 (0-18250) | | 3650 (0-29200) | |
| Pomace oil | 0 (0-500) | | 0 (0-520) | |
| Corn oil | 0 (0-3650) | | 0 (0-3650) | |
| Sunflower oil | 110 (0-3650) | | 0 (0-52080) | |
| Margarine | 0 (0-4380) | | 0 (0-4380) | |
| Sum of Group | 6360 | | 7300 | |
| **Cereals** | | | | |
| Corn | 0 (0-10950) | 10.91 | 0 (0-11250) | 4.75 |
| Mix | 0 (0-43800) | | 0 (0-21900) | |
| White wheat bread | 14600 (0-109500) | | 7300 (0-87600) | |
| Wholemeal bread | 480 (0-51100) | | 20 (0-73000) | |
| Maryland cookies | 168 (0-15330) | | 0 (0-10220) | |
| Sweet crackers | 0 (0-29200) | | 0 (0-10950) | |
| Chocolate biscuit | 10 (0-29200) | | 0 (0-10950) | |
| Bagel | 240 (0-41600) | | 480 (0-7300) | |
| Rice | 6240 (0-21900) | | 6240 (0-21900) | |
| Pasta | 6240 (0-21900) | | 6240 (0-21900) | |
| Sum of Group | 27978 | | 20280 | |
| **Chocolate** | | | | |
| Chocolate bun | 0 (0-21900) | 0.61 | 1560 (0-32850) | 0.7 |
| Chocolate | 1560 (0-9360) | | 1440 (0-21900) | |
| Sum of Group | 1560 | | 3000 | |
| **Legumes** | | | | |
| Lentil | 15600 (0-109500) | 11.70 | 15600 (0-31200) | 6.18 |

**Table 3. (Continued)**

| | Intake frequency | | | |
|---|---|---|---|---|
| | SURVEY 2013-2014 | | Survey 2005 | |
| Food group | Median: g/person/year (min-max) | Contribution (%) | Median: g/person/year (min-max) | Contribution (%) |
| Chickpea | 10800 (0-6800) | | 7200 (0-46800) | |
| White bean | 3600 (0-18000) | | 3600 (0-46800) | |
| Sum of Group | 30000 | | 26400 | |
| **Vegetables** | | | | |
| Chard | 2400 (0-73000) | 49.02 | 2000 (0-20800) | 64.11 |
| Spinach | 4800 (0-73000) | | 2400 (0-52000) | |
| Artichoke | 3200 (0-73000) | | 10400 (0-52000) | |
| Celery | 0 (0-73000) | | 0 (0-31200) | |
| Leek | 2400 (0-114400) | | 400 (0-41600) | |
| Eggplant | 4800 (0-31200) | | 7200 (0-41600) | |
| Marrow | 4800 (0-73000) | | 4800 (0-31200) | |
| Pumpkin | 50 (0-48000) | | 300 (0-15600) | |
| Onion | 15600 (0-109500) | | 10400 (0-109500) | |
| Champignon Mushroom | 7500 (0-31200) | | 3600 (0-23400) | |
| Cabbage | 200 (0-41600) | | 800 (0-73000) | |
| Cauliflower | 500 (0-41600) | | 400 (0-31200) | |
| Green asparagus | 180 (0-7800) | | 360 (0-10950) | |
| Pea | 1200 (0-36500) | | 1200 (0-36500) | |
| Lima bean | 2400 (0-20800) | | 2400 (0-73000) | |
| Green bean | 5200 (0-109500) | | 5200 (0-73000) | |
| Lettuce | 10400 (0-54750) | | 10400 (0-36500) | |
| Endive | 0 (0-18250) | | 100 (0-36500) | |
| Escarole | 300 (0-18250) | | 0 (0-36500) | |
| Roast potato | 7200 (0-109500) | | 7800 (0-54750) | |
| French fries | 600 (0-7800) | | 600 (0-36500) | |
| Cucumber | 3280 (0-146000) | | 2400 (0-146000) | |
| Red pepper | 4160 (0-87600) | | 4160 (0-29200) | |
| Green pepper | 4160 (0-58400) | | 1920 (0-29200) | |
| Mushroom | 2400 (0-53000) | | 800 (0-31200) | |
| Tomato | 28080 (0-219000) | | 33600 (0-219000) | |
| Carrot | 9360 (0-98550) | | 4680 (0-459360) | |
| Garlic | 520 (0-3650) | | 260 (0-1825) | |
| Sum of Group | 125690 | | 273750 | |
| **Fruits** | | | | |
| Orange | 15600 (0-272759) | 23.53 | 39000 (0-273750) | 21.58 |
| Grapefruit | 0 (0-7800) | | 0 (0-54750) | |
| Banana | 8320 (0-54750) | | 4160 (0-29200) | |
| Apple | 15600 (0-109500) | | 15600 (0-109500) | |
| Pear | 7800 (0-109500) | | 7800 (0-109500) | |

| | Intake frequency | | | |
|---|---|---|---|---|
| | Survey 2013-2014 | | Survey 2005 | |
| **Food group** | **Median: g/person/year (min-max)** | **Contribution (%)** | **Median: g/person/year (min-max)** | **Contribution (%)** |
| Strawberry | 550 (0-20800) | | 1200 (0-36500) | |
| Cherry | 600 (0-54730) | | 1200 (0-54750) | |
| Plum | 1125 (0-54750) | | 450 (0-54750) | |
| Peach | 4400 (0-73000) | | 7200 (0-73000) | |
| Apricot | 400 (0-219000) | | 800 (0-73000) | |
| Watermelon | 1250 (0-91250) | | 5000 (0-91250) | |
| Melon | 750 (0-33000) | | 3750 (0-91250) | |
| Kiwi | 3600 (0-73000) | | 3600 (0-36500) | |
| Grape | 350 (0-72800) | | 1960 (0-63875) | |
| Sum of Group | 60345 | | 92160 | |
| **Others** | | | | |
| Mayonnaise | 480 (0-7300) | 1.12 | 240 (0-20800) | 0.34 |
| Olives | 2400 (0-36500) | | 1200 (0-20800) | |
| Sum of Group | 2880 | | 1440 | |
| Total Median | 256428 | 100 | 427000 | 100 |

**Table 4. Plant sterols intake in 2013-2014 and 2005 surveys, expressed as mean mg PS/day and the contribution in percentage from each food group**

| Intake of plant sterols | | | | |
|---|---|---|---|---|
| **Food group** | Survey 2013-2014 | | Survey 2005 | |
| | **Mean: mg PS/day (min-max)** | **Contribution (%)** | **Mean: mg PS/day (min-max)** | **Contribution (%)** |
| **Nuts** | | | | |
| Almond | 1.95 (0-59.44) | 2.42 | 3.91 (0-59.44) | 3.87 |
| Cashew | 0.00 (0-9.00) | | 0.00 (0-9.00) | |
| Hazelnut | 0.07 (0-51.20) | | 0.042 (0-51.24) | |
| Peanut | 1.88 (0-244.47) | | 3.85 (0-58.52) | |
| Roasted corn | 0.00 (0-19.95) | | 0.00 (0-7.98) | |
| Nut | 1.73 (0-209.6) | | 1.73 (0-105.04) | |
| Sunflower seed | 0.47 (0-68.07) | | 0.019 (0-68.07) | |
| Pistachio | 0.53 (0-96.80) | | 0.08 (0-97.08) | |
| Sum of Group | 6.63 | | 10.90 | |
| **Oils and fats** | | | | |
| Refined olive oil | 16.80 (0-94.36) | 16.14 | 23.59 (0-165.13) | 17.58 |
| Virgin olive oil | 25.97 (0-129.85) | | 25.97 (0-207.76) | |
| Pomace oil | 0.00 (0-3.82) | | 0.00 (0-3.97) | |
| Corn oil | 0.00 (0-90.90) | | 0.00 (0-90.90) | |
| Sunflower oil | 1.48 (0-49.25) | | 0.00 (0-702.72) | |
| Margarine | 0.00 (0-26.04) | | 0.00 (0-26.04) | |
| Sum of Group | 44.26 | | 49.56 | |

**Table 4. (Continued)**

| Food group | Intake of plant sterols | | | |
|---|---|---|---|---|
| | Survey 2013-2014 | | Survey 2005 | |
| | Mean: mg PS/day (min-max) | Contribution (%) | Mean: mg PS/day (min-max) | Contribution (%) |
| **Cereals** | | | | |
| Corn | 0.00 (0-6.6) | 10.33 | 0.00 (0-6.78) | 6.85 |
| Mix | 0.00 (0-87.60) | | 0.00 (0-43.80) | |
| White wheat bread | 16.76 (0-125.70) | | 8.38 (0-100.56) | |
| Wholemeal bread | 0.92 (0-98) | | 0.04 (0-140) | |
| Maryland cookies | 0.31 (0-28.14) | | 0.00 (0-18.76) | |
| Sweet crackers | 0.00 (0-39.20) | | 0.00 (0-14.70) | |
| Chocolate biscuit | 0.02 (0-42) | | 0.00 (0-21) | |
| Bagel | 0.58 (0-100.30) | | 1.16 (0-17.60) | |
| Rice | 4.96 (0-17.40) | | 4.96 (0-17.40) | |
| Pasta | 4.79 (0-16.80) | | 4.79 (0-16.80) | |
| Sum of Group | 28.33 | | 19.32 | |
| **Chocolate** | | | | |
| Chocolate bun | 0.00 (0-19.80) | 1.96 | 0.00 (0-8.46) | 1.76 |
| Chocolate | 5.39 (0-113.40) | | 4.97 (0-75.6) | |
| Sum of Group | 5.39 | | 4.97 | |
| **Legumes** | | | | |
| Lentil | 50.13 (0-351.90) | 35.26 | 50.13 (0-100.27) | 30.04 |
| Chickpea | 35.86 (0-155.40) | | 23.91 (0-155.40) | |
| White bean | 10.66 (0-53.31) | | 10.66 (0-138.60) | |
| Sum of Group | 96.66 | | 4.70 | |
| **Vegetables** | | | | |
| Chard | 1.09 (0-33.2) | 19.27 | 0.91 (0-9.46) | 20.05 |
| Spinach | 2.14 (0-32.60) | | 1.07 (0-23.22) | |
| Artichoke | 4.25 (0-97.00) | | 13.82 (0-69.10) | |
| Celery | 0.00 (0-15.60) | | 0.00 (0-6.67) | |
| Leek | 0.77 (0-36.67) | | 0.13 (0-13.33) | |
| Eggplant | 0.78 (0-5.04) | | 1.16 (0-6.72) | |
| Marrow | 0.32 (0-4.80) | | 0.32 (0-2.05) | |
| Pumpkin | 0.02 (0-15.78) | | 0.10 (0-5.13) | |
| Onion | 3.08 (0-21.60) | | 2.05 (0-21.60) | |
| Champignon Mushroom | 3.70 (0-15.39) | | 1.78 (0-11.54) | |
| Cabbage | 0.15 (0-31.23) | | 0.60 (0-54.80) | |
| Cauliflower | 0.25 (0-20.86) | | 0.20 (0-15.64) | |
| Green asparagus | 0.05 (0-2.27) | | 0.10 (0-3.18) | |
| Pea | 4.44 (0-135.0) | | 4.44 (0-135.00) | |
| Lima bean | 8.15 (0-70.66) | | 8.15 (0-248.0) | |

| Intake of plant sterols | | | | |
|---|---|---|---|---|
| **Food group** | **Survey 2013-2014** | | **Survey 2005** | |
| | **Mean: mg PS/day (min-max)** | **Contribution (%)** | **Mean: mg PS/day (min-max)** | **Contribution (%)** |
| Green bean | 2.68 (0-56.40) | | 2.68 (0-37.60) | |
| Lettuce | 3.85 (0-20.25) | | 3.85 (0-13.50) | |
| Endive | 0.00 (0-8.45) | | 0.05 (0-16.90) | |
| Escarole | 0.15 (0-9.05) | | 0.00 (0-18.10) | |
| Roast potato | 0.85 (0-12.90) | | 0.92 (0-6.45) | |
| French fries | 0.39 (0-12.90) | | 0.39 (0-24.00) | |
| Cucumber | 0.63 (0-28.00) | | 0.46 (0-28.00) | |
| Red pepper | 0.19 (0-17.76) | | 0.84 (0-5.92) | |
| Green pepper | 1.07 (0-15.04) | | 0.49 (0-7.52) | |
| Mushroom | 1.18 (0-36.00) | | 0.39 (0-15.39) | |
| Tomato | 7.62 (0-59.40) | | 9.11 (0-59.40) | |
| Carrot | 4.77 (0-50.22) | | 2.38 (0-234.08) | |
| Garlic | 0.26 (0-1.82) | | 0.13 (0-0.91) | |
| Sum of Group | 52.82 | | 56.54 | |
| **Fruits** | | | | |
| Orange | 12.99 (0-228.09) | 11.28 | 32.48 (0-228) | 18.23 |
| Grapefruit | 0.00 (0-3.63) | | 0.00 (0-25.5) | |
| Banana | 4.58 (0-30.15) | | 2.29 (0-16.08) | |
| Apple | 6.84 (0-48.0) | | 6.84 (0-48) | |
| Pear | 2.35 (0-33.00) | | 2.35 (0-33) | |
| Strawberry | 0.17 (0-6.44() | | 0.37 (0-11.3) | |
| Cherry | 0.33 (0-30.15) | | 0.66 (0-30.15) | |
| Plum | 0.58 (0-28.35) | | 0.23 (0-28.35) | |
| Peach | 1.76 (0-29.20) | | 2.88 (0-29.2) | |
| Apricot | 0.17 (0-91.80) | | 0.33 (0-30.4) | |
| Watermelon | 0.15 (0-11.25) | | 0.62 (0-11.25) | |
| Melon | 0.09 (0-4.16) | | 0.47 (0-11.5) | |
| Kiwi | 0.70 (0-14.20) | | 0.70 (0-7.1) | |
| Grape | 0.19 (0-40.09) | | 1.16 (0-35.17) | |
| Sum of Group | 30.92 | | 51.39 | |
| **Others** | | | | |
| Mayonnaise | 6.36 (0-96.80) | 3.34 | 3.18 (0-96.80) | 1.62 |
| Olives | 2.79 (0-42.50) | | 1.40 (0-24.22) | |
| Sum of Group | 9.16 | | 4.58 | |
| **Total Mean** | 274.16 | 100 | 281.96 | 100 |
| **Total Median** | 466.69 | | 458.74 | |

In relation to PS intake from natural sources, on considering the mean mg/day consumed, most studies do not report values beyond 300 mg/day - the exception being the most recent study in Finland (Marttinen et al., 2014), with a slightly higher mean intake of 325 mg/day. On considering gender, men showed a higher PS intake, in general above 300 mg/day from natural sources, with the exception of Ireland and Sweden. In addition, and considering the different countries, the highest values for men were recorded in Finland (Marttinen et al.,

2014), with 363 mg/day, while the lowest were reported in Sweden (Klingberg et al., 2012) with 226 mg/day. In the case of women, PS intake was usually lower than 300 mg/day - the United Kingdom and Sweden being the countries with the highest and lowest PS intakes, respectively.

Finally, in these studies it can be seen that the main PS contributions from natural sources correspond to cereals, while dairy products are the main enriched source of PS in the diet. The results obtained in our study agree with those reported in the literature, and fall within the range of 211-325 mg/day, though a difference is the fact that the main PS contributor from natural food sources in our study corresponded to legumes.

## Conclusion

Based on the results obtained in the present study, it can be concluded that:

- The estimated mean PS intake in the studied resident population in the Valencian Community (Spain) for the period of 2013-2014 (274 mg/day) was very similar to the figure recorded in 2005 (282 mg/day).
- The main PS contribution from natural food sources corresponded to oils and fats, followed by nuts and legumes.
- Food frequency of intake was similar in both surveys (2013-2014 and 2005), vegetables being the most consumed foods, followed by fruits and legumes.
- The contribution to PS intake in the period 2013-2014 followed the order: legumes > vegetables > oils and fats; while in the survey of 2005 the ranking was legumes > vegetables > fruits.
- The estimation of PS intake in the most recent survey (2013-2014) yielded values very close to those reported in 2006 for the Spanish population (274.2 vs. 276 mg/day, respectively), and in good agreement with the range of values reported in different European studies (211-325 mg/day).

## Acknowledgments

This study has been financially supported by the Spanish Ministry of Economy and Competitiveness through Project AGL2012-39503-CO2-01 (CICYT-FEDER).

## References

Awad, A. B. & Fink, C. S. (2000). Phytosterols as anticancer dietary components: evidence and mechanism of action. *J Nutr*, *130*, 2127-2130.

Awad, A. B., Smith, A. J. & Fink, C. S. (2001). Plant sterols regulate rat vascular smooth muscle cell growth and prostacyclin release in culture., *Prostaglandins Leukot Essent Fatty Acids*, *64*, 323-330.

Awad, A. B., von Holtz, R. L., Cone, J. P., Fink, C. S. & Chen Y. C. (1998). β-Sitosterol inhibits the growth of HT-29 human colon cancer cells by activating the sphingomyelin cycle. *Anticancer Res*, *18*, 471-479.

Bacchetti, T., Masciangelo, S., Bicchiega, V., Bertoli, E. & Ferretti, G. (2011). Phytosterols, phytostanols and their esters: from natural to functional foods. *Med J Nutr Metab*, *4*, 165-172.

Bradford, P. G. & Awad, A. B. (2010). Modulation of signal transduction in cancer cells by phytosterols. *Biofactors*, *36*, 241-247.

Brüll, F., Ronald, P., Mensink. & Plat, J. (2009). Plant sterols: functional lipids in immune function and inflammation. *J Clin Lipidol*, *4*, 355-365.

Derdemezis, C. S., Filippatos, T. D., Mikhailidis, D. P., Mikhailidis, D. P. & Elisaf, M. S. (2010). Effects of plant sterols and stanols beyond low-density lipoprotein cholesterol lowering. *J Cardiovasc Pharmacol Ther*, *15*, 120-134.

EFSA, 2012. Scientific opinion on the substantiation of a health claim related to 3g/day plant sterols/stanols and lowering blood LDL-cholesterol and reduced risk of (coronary) heart disease pursuant to Article 19 of Regulation (EC) No 1924/2006. *EFSA J. 10*(5), 2693, 1-15.

Espín, J. C. & Tomás-Barberán, F. A. (2005). Constituyentes bioactivos no-nutricionales de alimentos de origen vegetal y su aplicación en alimentos funcionales. In M. Juárez, A. Olano. & F. Morais (Eds), Alimentos funcionales (135-137). Madrid, Spain: Fundación Española para la Ciencia Y Tecnología (FECYT).

European Commission Decision n. 2004/333/EC, of 31 March 2004 authorising the placing on the market of yellow fat spreads, salad dressings, milk type products, fermented milk type products, soya drinks and cheese type products with added phytosterols/phytostanols as novel foods or novel food ingredients under Regulation (EC) No 258/97 of the European Parliament and of the Council. *Off. J. Eur. Union*, *105*, 40-42.

Gabay, O., Sanchez, C., Salvat, C., Chevy, F., Breton, M., Nourissat, G., Wolf, C., Jacques, C. & Berenbaurn, F. (2010). Stigmasterol: a phytosterol with potential anti-osteoarthritic properties. *Osteoarthritis Cartilage*, *18*, 106-116.

García-Llatas, G. & Rodríguez-Estrada, M. T. (2011). Current and new insights on phytosterol oxides in plant sterol-enriched food. *J Chem Phys Lip*, *164*, 607–624.

Gylling, H., Plat J., Turley, S., Ginsberg, H. N., Ellegård, L., Jessup, W., Jones, P. J., Lütjohann, D., Maerz, W., Masana, L., Silbernagel, G., Staels, B., Borén, J., Catapano, A. L., De Backer, G., Deanfield, J., Descamps, O. S., Kovanen, P. T., Riccardi, G., Tokgözoglu L. & Chapman, M. J. (2014). Plant sterols and plant stanols in the management of dyslipidaemia and prevention of cardiovascular disease. *Atherosclerosis*, *232*, 346-360.

Hansel, B., Courie, R., Bayet, Y., Delestre, F. & Bruckert, E. (2011). Phytstérols et athérosclérose. *Red Med Internet*, *32*,124-129.

Hearty, A., Duffy, E. & Gibney, M. J. (2009). Intake estimates of naturally occurring phytosterols using deterministic and probabilistic methods in a representative Irish population. *Int J Food Sci Nutr*, *60*, 533-546.

Hearty, A., Duffy, E., Joyce, J., O'Connor, C. & Gibney, M. (2008). Phytosterol-enriched products on the Irish market: examination of intake and consumption patterns. *Public Health Nutr*, *12*, 51-58.

Jakulj. L., Mohammed, H., Van Dijk, T. H., Boer, T., Turner, S., Groen, A. K., Vissers, M. N. & Stroes, E. S. G. (2013). Plasma plant sterols serve as poor markers of cholesterol absorption in man. *J Lipid Res*, *54*, 1144-1150.

Jiménez-Escrig, A., Santos-Hidalgo, A. B. & Saura-Calixto, F. (2006). Common sources and estimated intake of plant sterols in the Spanish diet. *J Agric Food Chem*, *54*, 3462-3471.

Jones, P. J. H. & AbuMweis, S. S. (2009). Phytosterols as functional food ingredients: linkages to cardiovascular disease and cancer. *Curr Opin Clin Nutr Metab Care*, *12*, 147–151.

Klingberg, S., Andersson, H., Mulligan, A., Bhaniani, A., Welch, A., Bingham, S., Khaw, K. T., Andersson, S. & Ellegard, L. (2008). Food sources of plant sterols in the EPIC Norfolk population. *Eur J Clin Nutr*, *62*, 695-703.

Klingberg, S., Winsvist, A., Hallmans, G. & Johansson, I. (2012). Evaluation of plant sterol intake estimated with the northern Sweden FFQ. *Public Health Nutr*, *16*, 460-467.

Kuhlmann, K., Lindtner, O., Bauch, A., Ritter, G., Woerner, B. & Niemann, B. (2005). Simulation of prospective phytosterol intake in Germany by novel functional foods. *Br J Nutr*, *93*, 377-385.

Lama-Muñoz, A., Rodríguez-Gutiérrez, G., Rubio-Senent, F., Gómez-Carretero, A. & Fernández-Bolaños, J. (2011). New hydrothermal treatment of alperujo enhances the content of bioactive minor components in crude pomace olive oil. *J Agric Food Chem*, *59*, 1115-1123.

Marangoni, F. & Poli, A. (2010). Phytosterols and cardiovascular health. *Pharmacol Res*, *61*, 193-199.

Marttinen, M., Kosola, M., Ovaskainen, M-L., Mutanen, M. & Männisto, S. (2014). Plant sterol and stanol intake in Finland: a comparison between users and nonusers of plant sterol- and plant stanol-enriched foods. *Eur J Clin Nutr*, *68*, 587-591.

Normén, L., Brants, H., Voorrips, L., Andersson, H., Van den Brandt, P. & Goldbohm, A. (2001). Plant sterol intakes and colorectal cancer risk in the Netherlands Cohort Study on Diet and Cancer. *Am J Clin Nutr*, *74*, 141-148.

Normén, L., Bryngelsson, S., Johnsson, M., Evheden, P., Ellegard, L., Brants, H., Andersson, H. & Dutta, P. (2002). The phytosterol content of some cereal foods commonly consumed in Sweden and in the Netherlands. *J Food Comp Anal*, *15*, 693-704.

Normén, L., Ellegard, L., Brants, H., Dutta, P. & Andersson, H. (2007). A phytosterols database: fatty foods consumed in Sweden and the Netherlands. *J Food Comp Anal*, *20*, 193-201.

Normén, L., Johnsson, M., Andersson, H., Van Giameren, Y. & Dutta, P. (1999). Plant sterols in vegetables and fruits commonly consumed in Sweden. *Eur J Nutr*, *38*, 84-89.

Ostlund, R. E. (2002). Phytosterols in human nutrition. *Annu Rev Nutr*, *22*, 533-49.

Palou, A., Picó, C., Bonet, M. L., Serra, F., Rodríguez, A. M. & Ribot, J. (2005). El libro blanco de los esteroles vegetales en alimentación (2nd edition). Madrid, Spain: Instituto Flora.

Paniagua-Pérez, R., Madrigal-Bujaidar, E., Reyes-Cadena, S., Álvarez-González, I., Sánchez-Chapul, L., Pérez-Gallaga, J., Hernández, N., Flores-Mondragón, G. & Velasco, O. (2008). Cell protection induced by beta-sitosterol: inhibition of genotoxic damage, stimulation of lymphocyte production, and determination of its antioxidant capacity. *Arch Toxicol*, *82*, 615–622.

Rao, A. V. & Janezic, S. A. (1992). The role of dietary phytosterols in colon carcinogenesis. *Nutr Cancer*, *18*, 43-52.

Ras, R. T., Hiemstra, H., Lin, Y., Vermeer, M. A., Duchateau, G. S. M. J. E. & Trautwein, E. A. (2013). Consumption of plant sterol-enriched foods and effects on plasma plant sterol concentrations - A meta-analysis of randomized controlled studies. *Atherosclerosis*, *230*, 336-346.

Regulation 432/2012/EC of 16 May 2012 establishing a list of permitted health claims made on foods, other than those referring to the reduction of disease risk and to children's development and health. (2012). *Off. J. Eur. Union*, *L136*, 1-40.

Regulation 686/2014/EC of 20 June 2014 amending Regulations (EC) No 983/2009 and (EU) No 384/2010 as regards the conditions of use of certain health claims related to the lowering effect of plant sterols and plant stanols on blood LDL-cholesterol. (2014). *Off. J. Eur. Union*, *L182*, 27-30.

Regulation 983/2009/EC of 21 October 2009 on the authorisation and refusal of authorisation of certain health claims made on food and referring to the reduction of disease risk and to children's development and health. (2009). *Off. J. Eur. Union*, *L277*, 3-12.

Rozner, S. & Garti, N. (2006). The activity and absorption relationship of cholesterol and phytosterols. *Colloids Surf A Physicochem Eng Asp*, *282*, 435-456.

Rudkowska, I. (2010). Plant sterols and stanols for healthy ageing. *Maturitas*, *66*, 158–162.

Sioen, I., Matthys, C. H., Huybrechts, I., Van Camp, J. & De Henauw, S. (2011). Consumption of plant sterols in Belgium: estimated intakes and sources of naturally occurring plant sterols and β-carotene. *Br J Nutr*, *105*, 960-966.

Souci, S. W., Fachman, W. & Kraut, H. (2000). Food Composition and Nutrition Tables. Die Zusammensetzung der Lebensmittel Nährwert-Tabellen. La composition des aliments Tableaux des valeurs nutritive (6th edition). Stuttgart, Germany : Medpharm.

Valsta, L., Lemström, A., Ovaskainen, M. L., Lampi, A. M,, Toivo, J., Korhonen, T., & Piironen, V. (2004). Estimation of plant sterol and cholesterol intake in Finland: quality of new values and their effect on intake. *Br J Nutr*, *92*, 671-678.

Woyengo, T. A., Ramprasath, V. R. & Jones, P. J. H. (2009). Anticancer effects of phytosterols. *Eur J Clin Nutr*, *63*, 813-820.

# INDEX

## D



## E

## F



## G

## H



## I

## K



## L

## O



## P

## Q



## R

## S

## W



## Y



## Z